SONOCHEMISTRY:
Theory, Applications and Uses
of Ultrasound in Chemistry

ELLIS HORWOOD SERIES IN PHYSICAL CHEMISTRY

Series Editor: Professor T. J. KEMP, Department of Chemistry and Molecular Science, University of Warwick

Atherton, N.	Electron Spin Resonance Spectroscopy
Ball, M.C. & Strachan, A.N.	Chemistry and Reactivity of Solids
Buxton, G. V. & Salmon, G.A.	Pulse Radiolysis and its Applications in Chemistry
Cook, D.B.	Structures and Approximations for Electrons in Molecules
Coyle, J.D. & Horspool, W.	Organic Photochemistry
Cullis, C.F. & Hirschler, M.	Combustion and Air Pollution: Chemistry and Toxicology
Davies, P.B. & Russell, D.K.	Laser Magnetic Resonance
Devonshire, R.	Physical Photochemistry
Fadini, A. & Schnepel, F.-M.	Vibrational Spectroscopy
Harriman, A.	Inorganic Photochemistry
Horvath, A.L.	Handbook of Aqueous Electrolyte Solutions
Jankowska, H., Kwiatkowski, A. & Choma, J.	Active Carbon
Jaycock, M.J. & Parfitt, G.D.	Chemistry of Interfaces
Jaycock, M.J. & Parfitt, G.D.	Chemistry of Colloids
Ladd, M.F.C.	Structure and Bonding in Solid State Chemistry
Ladd, M.F.C.	Symmetry in Molecules and Crystals
Mason, T.J. & Lorimer, P.	Sonochemistry: Theory, Applications and Uses of Ultrasound in Chemistry
Milinchuk, V.K. & Tupikov, V.I.	Organic Radiation Chemistry Handbook
Mills, A. & Darwent, J.R. & Douglas, P.	Photochemical and Photoelectrical Conversion of Solar Energy
Navratil, O., Hala, J., Kopune, R. Leseticky, L., Macasek, F. & Mikulai, V.	Nuclear Chemistry
Paryjczak, T.	Gas Chromatography in Adsorption and Catalysis
Rest, A. & Dunkin, I.R.	Cryogenic Photochemistry
Sadlej, J.	Semi-empirical Methods in Quantum Chemistry
Snatzke, G., Ryback, G. & Slopes, P. M.	Optical Rotary Dispersion and Circular Dichroism
Southampton Electrochemistry Group	Instrumental Methods in Electrochemistry
Wan, J.K.S. & Depew, M.C.	Polarization and Magnetic Effects in Chemistry

SONOCHEMISTRY:
Theory, Applications and Uses of Ultrasound in Chemistry

TIMOTHY J. MASON, B.Sc., Ph.D., C.Chem., F.R.S.C
Principal Lecturer in Organic Chemistry
Coventry Polytechnic

and

J. PHILLIP LORIMER, B.Sc., Ph.D.
Senior Lecturer in Physical and Polymer Chemistry
Coventry Polytechnic

ELLIS HORWOOD LIMITED
Publishers · Chichester

Halsted Press: a division of
JOHN WILEY & SONS
New York · Chichester · Brisbane · Toronto

First published in 1988 by
ELLIS HORWOOD LIMITED
Market Cross House, Cooper Street,
Chichester, West Sussex, PO19 1EB, England
The publisher's colophon is reproduced from James Gillison's drawing of the ancient Market Cross, Chichester.

Distributors:

Australia and New Zealand:
JACARANDA WILEY LIMITED
GPO Box 859, Brisbane, Queensland 4001, Australia

Canada:
JOHN WILEY & SONS CANADA LIMITED
22 Worcester Road, Rexdale, Ontario, Canada

Europe and Africa:
JOHN WILEY & SONS LIMITED
Baffins Lane, Chichester, West Sussex, England

North and South America and the rest of the world:
Halsted Press: a division of
JOHN WILEY & SONS
605 Third Avenue, New York, NY 10158, USA

South-East Asia
JOHN WILEY & SONS (SEA) PTE LIMITED
37 Jalan Pemimpin # 05–04
Block B, Union Industrial Building, Singapore 2057

Indian Subcontinent
WILEY EASTERN LIMITED
4835/24 Ansari Road
Daryaganj, New Delhi 110002, India

© 1988 T. J. Mason and J. P. Lorimer/Ellis Horwood Limited

British Library Cataloguing in Publication Data
Mason, T. J. (Timothy James). *1946–*
Sonochemistry: Theory, Applications and Uses of Ultrasound in Chemistry
1. Chemistry. Use of ultrasonic waves
I. Title   II. Lorimer, J. P. (John Phillip)
542

Library of Congress CIP data available

ISBN 0–7458–0240–0 (Ellis Horwood Limited)
ISBN 0–470–21373–6 (Halsted Press)

Typeset in Times by Ellis Horwood Limited
Printed in Great Britain by Unwin Bros., Woking

Speakers at the First International Symposium on Sonochemistry. Held as part of the Royal Society of Chemistry Annual Congress, Warwick University, April 1986

FRONT ROW (left to right)

R. VERRALL, Department of Chemistry University of Sakatchewan, Sakatoon, Canada S7N OWO

Investigations into the physical principles of sonochemistry particularly by studies of sonoluminescence.

A. HENGLEIN, Bereich Strahlenchemie, Hanm Meitner–Institut, Postfach 39 01 28, D 1000 Berlin 39, Glienicker Strasse 100

Wide ranging studies including fundamental investigations of cavitation phenomena in both aqueous and non-aqueous media. A pioneer of sonochemistry with published work on the uses of ultrasound in polymer chemistry dating back to the 1950s.

T. J. MASON, School of Chemistry, Coventry Polytechnic, Coventry CV1 5FB, United Kingdom

Part of the Coventry Sonochemistry group whose wide ranging studies include fundamental investigations of the effect of cavitation on reaction kinetics, organic and organometallic synthesis and polymer chemistry. Particular interests in physical, organic and polymer chemistry applications.

P. BOUDJOUK, Department of Chemistry, North Dakota State University Fargo, North Dakota 58105–5516, USA

Studies mainly concerned with the effects of ultrasound in synthetic chemistry involving organometallic compounds.

K. S. SUSLICK, School of Chemical Sciences, University of Illinois at Urbana Champaign, Mailroom 296 Roger Adams Laboratory, 1209 West California Street, Urbana, Illionois 61801, USA

Wide ranging studies including the underlying physical principles of cavitation phenomena and their application in organometallic chemistry and catalysis.

C. DUPUY, L.E.D.S.S., Batiment 52, Chimie Recherche, Universite Scientifique et Medical de Grenoble, St Martin D'Heres Cedex, BP 68 38402, Grenoble, France

Part of the research group of J.-L. Luche

SECOND ROW (left to right)

J. LINDLEY, School of Chemistry, Coventry Polytechnic, Coventry CV1 5FB, UK

Part of the Coventry Sonochemistry group whose wide ranging studies include fundamental investigations of the effect of cavitation on reaction kinetics, organic and organometallic synthesis and polymer chemistry. Particular interests in applications to inorganic and organometallic chemistry.

P. RIESZ, Department of Health & Human Services, Public Health Service, Building 10, Room B1B50 National Institutes of Health National Cancer Institute, Bethesda, Maryland 20205, USA

Main interests lie in the generation of free radicals on sonolysis of aqueous media. This work linked to the question of any possible harmful effects arising from the medical uses of ultrasound.

T. J. LEWIS, Department of Electronic Engineering, University College of North Wales, Dean Street, Bangor, Gwynedd LL57 1UT, UK

The uses of ultrasonic attenuation for the elucidation of molecular structure of materials of biological significane.

J. PERKINS, Sonic Systems, The Old Bakery, 18 Silverless Street, Marlbrough, Wiltshire SN8 1JQ, UK

An ultrasonic engineering consultant with particular expertise in the design of equipment suitable for sonochemistry.

J. EINHORN, L.E.D.S.S., Batiment 52, Chimie Recherche, Universite Scientifique et Medical de Grenoble, St Martin D'Heres Cedex, BP 68 38402, Grenoble, France

Part of the research group of J.-L. Luche

P. KRUUS, Department of Chemistry, Carleton University, Ottawa K1S 5B6, Canada

Particularly interested in ultrasonically initiated polymerisation reactions.

BACK ROW (left to right)

R. S. DAVIDSON, Department of Chemistry, City University, Northampton Square, London EC1V 0HB, UK

Interests in the physical effects of ultrasound, radical generation and organic synthesis.

J-L. LUCHE, L.E.D.S.S., Batiment 52, Chimie Recherche, Universite Scientifique et Medical de Grenoble, St Martin D'Heres Cedex, BP 68 38402, Grenoble, France

Main research interests in organic synthesis using organometallic intermediates. some of the work has led to a widening of the application of a number of older traditional methodologies, e.g. the Barbier reaction.

B. PUGIN, CIBA–CEIGY, K–1055–616, CH – 4002 – Basle, Switzerland

Interests lie in sonochemical reactor design and synthesis involving organometallics.

J. P. LORIMER, School of Chemistry, Coventry Polytechnic, Coventry CV1 5FB, UK

Part of the Coventry Sonochemistry group whose wide ranging studies include fundamental investigations of the effect of cavitation on reaction kinetics, organic and organometallic synthesis and polymer chemistry. Particular interests in physical and polymer applications.

Table of contents

Preface

I suppose it was a bit of an accident of fate which placed the two of us (an organic chemist — TJM, and a physical chemist — JPL) together in initial studies on sonochemistry and eventually led to this textbook.

In 1975 we were both relieved to find ourselves in permanent employment at the then Lanchester Polytechnic after our separate careers through the University system. Perhaps one advantage that the Polytechnic had over Universities at that time was that our department was actually Chemistry and Metallurgy. From a cloistered upbringing as a chemist it would have been difficult to predict the way in which the initial germ of an idea came from rubbing shoulders with metallurgists. it was purely by chance some time after my arrival that I happened to be walking through a metallurgy laboratory and saw an ultrasonic bath being used to clean metal samples. The process intrigued me for I could see that the ultrasonic bath was producing a large amount of energy as evidenced by the distrubance of the water with which it was filled. It occurred to me that this was perhaps a form of energy which might be employed to influence chemical reactivity — particularly solvolysis reactions which had been the object of my research for many years. And so I contrived to borrow the bath and spent a few days investigating the possibilities. The initial results were puzzling but sharing an office with JPL meant that it very quickly became apparent that the combined approach to this problem by chemists of different disciplines was vital. Neither of us had heard of using ultrasound as a source of energy to promote chemical reactivity. It was not a topic mentioned in undergraduate chemistry courses and neither was it something we had come across in the current chemical literature. And so there we were, confronting what was apparently a brand new research field, and with no sources of information to hand.

And so it started, some twelve years ago. Our first publication in 1980 was a Chemical Communication in which was reported small (up to two-fold) enhancements in the hydrolysis rate of 2-chloro-2-methylpropane. Progress was very slow initially although we kept going on solvolysis reactions. It is really only over the last five years that the topic has really begun to take off. In 1986 we were involved in organising the first ever international conference on Sonochemistry at Warwick University. *The Times* newspaper of 14 April 1986 wrote of our conference:

A new industrial revolution is in the offing, replacing traditional manufacturing of plastics, detergents, pharmaceuticals and agricultural chemicals. They will be safer processes because they avoid the high temperatures and high pressures of present methods. And they will be cheaper because they use less fuel than that needed to provide the heat energy for stimulating the chemical reaction. The way of injecting the energy at ordinary room temperatures depends on discoveries made in a new branch of science known as sonochemistry.

Sonochemistry is a term that has now started to appear regularly in the chemical literature in the context of its use to improve reaction rates and/or product yields. Our research team at Coventry has grown and we are fortunate to have won a premier position in the UK. In 1987 a Royal Society of Chemistry Subject Group devoted to Sonochemistry was formed which has three members of our research group serving on its committee. The first residential course was mounted in 1988 and attracted delegates from France, Israel, Japan, the Netherlands, Sweden, USA and Yugoslavia. It is because we have experienced the broad appeal of the topic that we decided to embark on writing this book.

The current wide availability of ultrasonic equipment facilitates a chemist's practical introduction to sonochemistry. Yet to make full use of the driving force behind sonochemically modified reactions — cavitation — it really is necessary to appreciate some of the background theory. The problem is that the majority of treatises on cavitation are written by physicists and mathematicians and, perhaps because of this, they are somewhat indigestible for chemists. In our book we have tried to keep theory at a level which is instructive but not overpowering.

It is never a good idea for any scientist to be overly insular about his own topic. As chemists we have a tendency to stick to our own discipline, even to our own branch of chemistry. The joy of sonochemistry is that it has applications across almost the whole breadth of chemistry. For this reason we decided to include a chapter on ultrasound in polymer chemistry together with the more 'orthodox' synthetic and physical sonochemistry. The chapter on equipment is gleaned from many sources — a potential sonochemist will find it impossible to get this type of information from manufacturers.

The use of ultrasound as a diagnostic tool in chemistry derives from the detailed and extensive work on non-destructive testing and medical imaging. We considered this to be an essential topic for our text even though ultrasonic irradiation at the frequencies employed for analysis cause no chemical changes in a system.

In the final chapter we have attempted to show how widespread are the existing uses of ultrasound. It is important that the industrialist recognises that the 'novelty' of using ultrasound in chemistry is not something to be feared — it is a common enough energy source in processing applications.

We are indebted to John Perkins, an ultrasonic engineer, who has always been most helpful when we have needed advice on equipment, to Kerry Ultrasonics and Roth Scientific, two companies who have assisted us throughout our research. We also thank our wives for their forbearance over the last year when most of our 'spare' time has been devoted to this book.

1

Ultrasonics

1.1 INTRODUCTION

If you were asked what you knew about ultrasound you would almost certainly start with the fact that it is used in animal communications (e.g. bat navigation and dog whistles). You might then recall that ultrasound is used in medicine for fetal imaging, in underwater range finding (SONAR) or in the non-destructive testing of metals for flaws. For a chemist, however, sound would probably not be the first form of energy that would be considered for the excitation of a chemical reaction. Indeed up to a few years ago the use of ultrasound in chemistry was something of a curiosity and the practising chemist could have been forgiven for not having met the concept. To increase chemical reactivity one would probably turn towards heat, pressure, light or the use of a catalyst. And yet, if one stops for a second to consider what is involved in the transmission of a sound wave through a medium it is perhaps surprising that for so many years sound was not considered as a potential source of enhancement of chemical reactivity. The only exception to this being the green-fingered chemist who, in the privacy of his own laboratory, talks, sings or even shouts at his reaction. After all, sound is transmitted through a medium as a pressure wave and the mere act of transmission must cause some excitation in the medium in the form of enhanced molecular motion.

The basis for the present-day generation of ultrasound was established as far back as 1880 with the discovery of the piezoelectric effect and its inverse by the Curies [1,2]. Most modern ultrasonic devices rely on transducers which use the inverse effect, i.e. the production of a change in dimension of certain materials by the application of an electrical potential across opposite faces (see Chapter 7). If the potential is alternated at high frequencies the crystal converts the electrical to mechanical (sound) energy — rather like a loudspeaker. At sufficiently high alternating potential ultrasound will be generated. The earliest form of ultrasonic transducer, however, was a whistle developed by Galton in 1883 to investigate the threshold frequency of human hearing [3].

The first commercial application of ultrasonics did not appear until 1917 with Langevin's echo-sounding technique for the estimation of depths of water. Langevin's discovery was the direct result of an idea which arose in suggestions generated by a competition organised in 1912 to find a method of detecting icebergs in the open sea and so avoid any repetition of the disaster which befell the Titanic. The early 'echo sounder' simply sent a pulse of ultrasound from the keel of a boat to the bottom of the sea from which it was reflected back to a detector also on the keel. For sound waves, since the distance travelled through a medium = 1/2 × time × velocity (and the velocity of sound in seawater is accurately known) the distance to the bottom could be gauged from the time taken for the signal to return to the boat. If some foreign object (e.g. a submarine) were to come between the boat and the bottom of the seabed an echo would be produced from this in advance of the bottom echo. This system was very important to the Allied Submarine Detection Investigation Committee during the war and became popularly known by the acronym ASDIC. Later developments resulted in the system known as SONAR (SOund Navigation And Ranging) which allowed the surrounding sea to be scanned. As an example of its efficiency, using modern SONAR it is possible to locate a small fish only 35 cm in length at a depth of 500 metres. The original ASDIC system predated the corresponding RAdio Detection And Ranging system (RADAR) by 30 years.

Essentially all imaging from medical ultrasound to non-destructive testing relies upon the same pulse-echo type of approach but with considerably refined electronic hardware. The refinements enable the equipment not only to detect reflections of the sound wave from the hard, metallic surface of a submarine in water but also much more subtle changes in the media through which sound passes (e.g. those between different tissue structures in the body). It is high-frequency ultrasound (in the range 2 to 10 MHz) which is used primarily in this type of application because by using these much shorter wavelengths it is possible to detect much smaller areas of phase change, i.e. give better 'definition'.

The chemical applications of high-frequency ultrasound are concerned essentially with measurements of the degree to which the sound is absorbed as it passes through a medium. This effect, known as attenuation, will be discussed in detail in Chapter 2 (section 2.3).

It is only since 1945 with the increased understanding of the phenomenon of cavitation, together with significant developments made in electronic circuitry and transducer design (i.e. devices which convert electrical to mechanical signals and vice versa), that a rapid expansion in the application of power ultrasound to chemical processes (Sonochemistry) has occurred. Power ultrasound affects chemical reactivity through cavitation. Cavitation as a phenomenon was first reported in 1895 by Sir John Thornycroft and Sidney Barnaby [4]. This was the result of investigations into the inexplicably poor performance of a newly built destroyer HMS *Daring*. Her top speed was well below specifications and the problem was traced to the propeller blades which were incorrectly set and therefore not generating sufficient thrust. The rapid motion of the blades through water was found to tear the water structure apart by virtue of simply mechanical action. The result of this was the production of what are now called cavitation bubbles.

1.2 SOUND RANGES

The range of human hearing is from about 16 Hz to 16 kHz (with middle C at 261 Hz and the meadow grasshopper operating at around 7 kHz). Ultrasound is the name given to sound waves having frequencies higher than those to which the human ear can respond (i.e. >16 kHz). The upper limit of ultrasonic frequency is one which is not sharply defined but is usually taken to be 5 MHz for gases and 500 MHz for liquids and solids. The uses of ultrasound within this large frequency range may be divided broadly into two areas. The first area involves low amplitude (higher frequency) propagation, which is concerned with the effect of the medium on the wave and is commonly referred to as *low power* or *high frequency ultrasound*. Typically, low amplitude waves are used to measure the velocity and absorption coefficient of the wave in a medium in the 2 to 10 MHz range. It is used in medical scanning, chemical analysis and the study of relaxation phenomena. The second area involves high energy (low frequency) waves known as *power ultrasound* between 20 and 100 kHz which is used for cleaning, plastic welding and, more recently, to effect chemical reactivity (see Fig. 1.1).

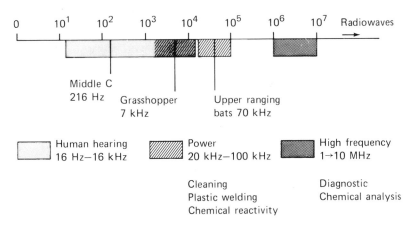

Fig. 1.1 — Sound frequencies (cps/Hz).

1.3 SOME CURRENT INDUSTRIAL USES OF ULTRASOUND

Before we discuss how sound energy affects chemical reactivity it would be instructive to explore some of the broader applications of the uses of ultrasound in industry and medicine [5]. Some of these have been in existence for many years and a range of such are shown in Table 1.1. Two of these applications have provided the direct antecedents of the types of equipment now commonly used for sonochemistry, namely ultrasonic welders and cleaning baths.

1.3.1 Ultrasonic Welding

A large proportion of ultrasonic equipment currently in industry is involved in welding or riveting plastic mouldings for the consumer market [6]. The equipment

Table 1.1 — Some industrial uses of ultrasound

Field	Application
Biology, biochemistry	Homogenisation and cell disruption: Power ultrasound is used to rupture cell walls in order to release contents for further studies.
Engineering	Ultrasound has been used to assist drilling, grinding and cutting. It is particularly useful for processing hard, brittle materials, e.g. glass, ceramics. Other uses of power ultrasound are welding (both plastics and metals) and metal tube drawing.
	High-frequency (MHz) ultrasound is used in non-destructive testing of materials and flaw detection.
Dentistry	For both cleaning and drilling of teeth.
Geography, geology	Pulse/echo techniques are used in the location of mineral and oil deposits and in depth gauges for seas and oceans. Echo ranging at sea has been used for many years (SONAR).
Industrial	Pigments and solids can be easily dispersed in paint, inks and resins. Engineering articles are often cleaned and degreased by immersion in ultrasonic baths. Two less widely used applications are acoustic filtration and ultrasound drying.
Medicine	Ultrasound imaging (2–10 MHz) is used, particularly in obstetrics, for observing the fetus and for guiding subcutaneous surgical implements. In physiotherapy lower frequencies (20–50 kHz) are used in the treatment of muscle strains.
Plastics and polymers	The welding of thermoplastics is effectively achieved using power ultrasound. The initiation of polymerisation and polymer degradation are also affected.
	Cure rates of resins and their composition can be measured with high-frequency ultrasound.

consists of a generator producing an alternating frequency of around 20 000 Hz (Hz is the abbreviation for Hertz, i.e. cycles per second) which feeds a transducer (normally piezoelectric or magnetostrictive — see Chapter 7), which converts the electrical energy into mechanical energy. A shaped tool or horn transmits (and amplifies) the vibrating motion to a shaped die pressing together the two pieces of material to be welded (Fig. 1.2). A useful analogy for the action of an ultrasonic welder is the pneumatic road drill, only with ultrasonics the frequency is higher and just above normal hearing, and the vibrational amplitude is much smaller, typically 50–100 micrometres (or 0.001 0.002 in) rather than several inches.

Fig. 1.2 — Ultrasonic welder. (Picture by courtesy of Kerry Ultrasonics Ltd, Hunting Gate Wilbury Way, Hitchin, Hertfordshire, UK.)

Ultrasonic welding is generally used for the more rigid amorphous types of thermoplastic. These are most suitable for welding because the ultrasonic vibration can travel through the bulk plastic of the component to the joint. The more flexible and crystalline the plastic material, the more readily it will absorb the vibration energy as it travels through the material, i.e. the 'sound' is attenuated (or deadened) rapidly as it passes through. This type of plastic is generally only welded in the form of thin sections of sheet or film. For plastic welding it is particularly important that the vibrational energy is transmitted only to the joint and not the body of the material, since any warming of the bulk material may lead to a release of internal moulding stresses and produce distortion.

Thermoplastics have two properties which make them particularly suited to ultrasonic welding (a) low thermal conductivity and (b) melting or softening temperatures of between 100 and 200°C. As soon as the ultrasonic power is switched off the substrate or bulk material becomes a heat-sink, giving rapid cooling of the welded joint. When the more traditional conductive heating is used for welding, however, the thermal gradient has to be reversed before cooling occurs, leading to long heating/cooling process cycles. Another major advantage of the use of ultra-sound is the high joint strength of the weld, reaching 90–98% of the material strength. Indeed, test samples usually break in the body of the material and not at the weld itself.

To emphasise the advantages of ultrasonic welding compared with the more traditional approach several examples can be quoted. For aesthetic reasons it is important to avoid distortions in spectacle frames and this has been achieved by ultrasonically welding the metal hinge joint into the plastic frame. Other examples are to be found in the automotive industry where, for example, the car rear reflector cluster will normally have its plastic lens welded into place and the threaded brass insert may well have been inserted by ultrasound. The 13 A mains plug often has the brass threaded nut ultrasonically inserted into the lid; the outer knurling is trapped in place by the plastic melted by the vibrational energy. Emulsion paint cans rust and discolour the paint, consequently several firms now use plastic cans to avoid this problem. The rim which takes the snap-on lid is welded ultrasonically to the body in an automated high-speed canning line.

Ultrasonic welding is not only restricted to plastics but can also be used for metals. In this case the ultrasonic motion must be lateral rather than vertical so that frictional heating is induced between the surfaces. One typical application is in the welding of aluminium which is difficult by normal methods because of its tenacious oxide. With ultrasonic metal welding — a form of low temperature diffusion welding — the oxide layer is easily broken up and adsorbed within the metal surrounding the weld. Welding by lateral vibrational movement is readily achieved using this technique without the formation of brittle intermetallic compounds. Ultrasonic metal welding, like plastic welding, permits the very delicate joining of components. In flute manufacture it is clearly important to avoid distortion when attaching pillars to the flute body. The well known musical instrument makers Boosey and Hawkes largely eliminated the problem by using an ultrasonic stud-welder for the process thus avoiding the heat normally associated with hard soldering.

1.3.2 Ultrasonic cleaning

Ultrasonic cleaning is another major application for power ultrasound. It is now such a well established technique that laboratories without access to an ultrasonic cleaning bath are in a minority. It is important to recognise the historical significance of the development of ultrasonic cleaning bath technology on the growth of sonochemistry because the use of ultrasonic cleaners is probably the first method to which the chemist will turn when starting sonochemistry research.

Although the laboratory ultrasonic cleaning bath is familiar to the chemist, the industrial applications of such cleaning are perhaps less well known. Yet it is developments in industrial cleaning which have made it possible to consider large-scale chemical reactions since the larger the batch chemical process the bigger will be the size requirement of the bath. Ultrasonic cleaning can be both delicately applied and used for very large items. Thus microcomponents for computer applications can be cleaned under clean room conditions. On a slightly larger scale valves for pressurised gas cylinders can be purged of all cutting oil residues and swarf (essential in preventing an explosion hazard in handling oxygen from gas cylinders). This type of cleaning can be achieved most effectively using sonicated chlorinated solvents.

Other applications include such diverse subjects as the cleaning of engine blocks, jewellery, medical instruments and the removal of contaminants and water resistant marker-pen ink used as guidelines for cutting in crystal glass manufacturing processes. It is when we turn to much bigger cleaning problems that we find that the equipment built for such specialist applications becomes ideal for large-scale chemical reactions.

One of the most publicised events in Britain in 1982 was the raising of the Tudor warship the *Mary Rose*. The ship had lain on the Solent seabed, buried by preserving silt, for 437 years and ultrasound was used in two ways to assist in this historic 'rescue' [7]. The initial location of the wreck was greatly assisted by SONAR. Subsequently, following her recovery, strenuous efforts were made by the *Mary Rose* Trust to preserve the 17,000 or so artefacts recovered from the wreck. These included hundreds of bows, arrows, archery equipment, musical instruments and personal objects such as clothes, footwear, coins and so forth. The prospect of cleaning and preserving this vast amount of material by laborious soaking and rinsing had proved daunting even before the *Mary Rose* itself broke the surface of the waves. A British firm, Kerry Ultrasonics, helped the trust by supplying a specially constructed ultrasonic cleaning tank presented to the *Mary Rose* Trust in May 1983 (Fig 1.3). The

Fig. 1.3 — Large-scale ultrasonic cleaning bath as used by the 'Mary Rose Trust'. (Picture by courtesy of Kerry Ultrasonics Ltd, Hunting Gate Wilbury Way, Hitchin, Hertfordshire, UK.)

bath was large enough to accommodate items as bulky as wooden gun carriages but could as easily be used for smaller items. One of the major problems in preserving artefacts constructed of organic materials — wood and leather — is that they often have become impregnated with iron deposits. These deposits have to be removed to prevent further deterioration, since iron staining and rust block the pores of the organic materials, preventing effective penetration of the chemical preservatives. Ultrasonic cleaning is very effective in removing such deposits and can, at a later stage, be used to improve preservative penetration.

For very large baths such as these it is necessary to use a great number of transducers attached to the base. In Fig. 1.4 the array of transducers attached to the

Fig. 1.4 — View of underside of a large (12×3 feet base) ultrasonic tank covered with early type magnetostrictive transducers rated at 15 kW. (Picture by courtesy of Lewis Corporation, 324 Christian Street, Oxford, Connecticut, USA.)

12′ x 3′ base of a Lewis Corporation bath is shown. In this example an older type of magnetostrictive transducer is used giving a total power of 15 kW.

Many complex (and expensive) multibath cleaning, rinsing and drying systems can be found in high volume production lines. For the laboratory, however, a single small cleaning bath can be obtained for no more than a few hundred pounds.

1.3.3 Ultrasound in biology

The primary biological use of ultrasound is for the disruption of cell walls to release the contents for *in vitro* studies. The type of apparatus used for this purpose is a direct descendent of the welding equipment described above. A horn resonating at 20 kHz is dipped into a suspension of biological cellular material. This process depends upon the efficiency with which the cell wall can be disrupted by the 'ultrasound' to release its cellular contents without at the same time destroying them. This is much more difficult than it appears on the surface. The problem is that most simple one-cell organisms have an exceedingly tough cell wall which is only a few microns in diameter, and similar in density to the medium that surrounds it. The protein and nucleic acid components contained within the cell are large macromolecules, easily denatured by extreme conditions of temperature or oxidation. Sonication of an aqueous suspension of cellular material efficiently disrupts the wall but avoids extensive destruction of the cellular components.

The breakage of the outer cell wall by ultrasound takes place by what has been described as a machine-gun-like phenomenon whereby extremely small cavitation bubbles, driven at very high speed from the probe tip of the sonic device, actually penetrate through the cell wall thereby disrupting it. The effective area for this penetration is close to the probe tip and the cell must be kept within the effective area of the probe for a sufficient length of time so that disruption takes place. A delicate balance must therefore be struck between the power of the probe and the disruption rate since power ultrasound, with its associated cavitational collapse energy and bulk heating effect, can denature the contents of the cell once released. Indeed, for this type of usage it is important to keep the cell sample cool during sonication.

1.3.4   Medical ultrasound

Diagnostic uses
Unlike the applications referred to above diagnostic medical scanners (and NDT equipment) work at very low power levels (milliwatts) on a pulsed echo system. As its name suggests, diagnostic (or high frequency) ultrasound provides a non-invasive technique for scanning the human body and can be considered to be complementary to X-ray and n.m.r. methods. One of the most publicised uses of diagnostic ultrasound has been in fetal imaging [8]. Another major usage is as a continuous and non-hazardous method of visualising the position of needles and other surgical implements as they are used within the body.

The origins of acoustic sensing in medical diagnosis go back a long way. Physicians have always used the direct approach of listening for the sounds generated from the heart and lungs but in the 18th century a new technique was introduced known as 'percussion'. This involved the physician tapping on the surface of the skin of a patient and listening for the resonant note emitted by the air-filled or liquid filled spaces within the body. Practice in the method allows diagnosis of a range of medical conditions.

The present day refined 'percussion' techniques involving pulsed ultrasound are derived from instruments originally designed for non-destructive testing. In the medical use of ultrasound, high-frequency acoustic pulses of short duration are propagated in more or less straight lines and are scattered as they encounter various physical gradients or boundaries within tissues. 'Back-scatter', in which the pulses are reflected 180°, results in a series of echoes which can be detected at the skin surface. The lapse of time between launching and receiving a pulse is proportional to the depth of the reflecting tissue (the average sound velocity in body tissues is about

1.54 km s^{-1}). Likewise, the time lapse between echoes corresponds to the space between structures that are reflecting the pulses. Signals can also vary in strength, or amplitude, depending on the specific features of the reflecting structures, such as their composition, compressibility, and density.

An ultrasound image is synthesised line by line, each line of the image representing the echo pattern of an individual acoustic pulse. The result is a spatial map of phase boundaries within the tissues. Signal amplitude is portrayed in a gray scale of tonal gradations, with the strongest reflected signals shown as intense white and the absence of a signal as black. Today's devices use acoustic pulses with an average frequency in the range of 3-10 MHz; pulse wavelengths and the corresponding resolution features of the image are equal to or less than 1 mm.

Clearly if faint echoes are received from changes in tissue within the body a small medical probe will be easily detected as an intense white image and hence ultrasound can be used in delicate operations (e.g. on the unborn fetus) to visualise and guide microprobes.

Medical scanners have an array of transducers in each probe and are capable of providing thousands of measurements of the small changes in the velocity of sound as it passes through various phase changes within the body. This information is fed to computers which create a television picture of the internal tissues of the human body.

Medical uses of power ultrasound

One of the first applications of ultrasound in medicine was the so-called ultrasonic massage introduced in Germany before the Second World War. This was introduced as a substitute for the hands of the masseur in patients who had suffered from fractures and similar injuries. Rubbing movements are capable of improving the circulation very considerably and help also to break down adhesions between muscles and their sheaths which limit the range of movement.

It was thought that the same effects would be obtained by the mechanical movement of the tissues with plane wave ultrasonic irradiation. This was generated by flat quartz crystals, the earthed surfaces of which could be applied directly to the skin. The use of ultrasound for the treatment of sporting injuries, particularly strains and 'tennis elbow' is now commonplace as equipment for this purpose is readily commercially available to the physiotherapist.

More recently power ultrasound has been used for the treatment of kidney stones. The method is simple and relies on the mechanical effect of ultrasound to break up the stones into particles small enough to be excreted normally. This rapid, non-invasive and non-chemical treatment is clearly a real boon to the patient suffering from this painful complaint.

The cleaning of medical instruments

The cleaning of instruments such as scalpels, forceps, etc., must be carried out prior to the sterilisation process to which the instruments are ultimately subjected. Sterilisation refers to the freeing of an object from all living organisms, including bacteria, and in surgery and medicine the sterilisation of instruments is extremely important in order to prevent infection. For surgical instruments the traditional method of cleaning is by hand scrubbing. This is a very time-consuming task and, moreover, the method is unsatisfactory since it has proved almost impossible to hand clean the inaccessible surfaces of modern surgical instruments. Ultrasonic baths are now used for the cleaning of instruments used in surgery and medicine providing a most efficient and rapid process.

1.3.5   Non-destructive testing (NDT)
Non-destructive testing was the precursor to medical scanning and most of the information given in the section above applies to this technique. Little more need be written about NDT except to emphasise the speed and accuracy with which a flaw detector can be used for a whole range of different materials.

1.3.6   Engineering applications
Ultrasonic machining (USM)
The development of ultrasonic machining processes was necessitated by the increasing usage of hard, brittle materials and the need to machine them effectively and accurately. USM is being used currently to machine carbides, stainless steels, ceramics and glass.

The ultrasonic machining process is performed by a cutting tool normally operating around 20 kHz in an abrasive slurry (containing silicon carbide, boron carbide or alumina). The shape of the tool corresponds to the shape produced in the workpiece. The first patent for such a machine was granted in England as far back as 1945. Ultrasound can also be used as a supplement to traditional machining giving greatly increased efficiencies.

Abrasive jet machining (AJM)
Abrasive jet machining differs from conventional sand-blasting in that the abrasive is much finer and the cutting action is more carefully controlled. AJM can be used to cut hard, brittle materials (germanium, silicon, mica, glass and ceramics) in a wide variety of cutting, deburring and cleaning operations. The process is free from chatter and vibration at the working face since the tool is not in contact with the workpiece.

Drilling and Cutting
The aerospace industry uses ultrasonic drilling of carbide turbine blades and other hard, brittle and friable materials. They also use tungsten carbide ultrasonic cutting blades to cut complex carbon-fibre shapes out of fibre sheets before the fibre is encapsulated in epoxy resin. With conventional techniques it is difficult to cut these tough materials quickly on a continuous basis and maintain a high quality finish.

Another use is in the production of three-dimensional engraved glass which can easily be produced by a shaped ultrasonic tool using an abrasive slurry.

Metal tube drawing
Remarkable improvements in the cold-drawing of metal tubing has been achieved when the die to be used is subjected to radial ultrasonic vibrations of 20 kHz. The technique has been applied to stainless steel and gives much larger reductions in size per pass, greatly reduced draw pressures and a better finish than that produced using traditional methods.

1.3.7 Dentistry

The connection between the use of ultrasound for the machining and drilling of hard materials and dentistry is obvious. In the dentist's surgery a new instrument which uses ultrasonics as its mode of action is now in common usage. Although these instruments were first produced back in the late fifties, technical improvements have seen their adaptation for an increasingly wide range of dental applications. Essentially it is a small ultrasonic probe operating at 25 kHz with attachments which allow it to be used for cleaning and polishing teeth, descaling, drilling and even root canal work.

For simple cleaning the dentist will use the instrument in a mode such that the vibrating tip is used to sonicate a fine spray of air, water and specially prepared sodium carbonate. Such a spray when precisely projected against the tooth surface has been found to have a cleaning/polishing effect which is far less abrasive towards tooth enamel than the more traditional grinding with pumice stone.

With a selection of different hook-shaped metal tools the same implement can be used for de-scaling making use of the 25 000 times per second push-pull motion of the resonating tip. If a straight diamond-tipped drill attachment is used in conjunction with an abrasive slurry containing alumina powder then the same motion gives efficient drilling.

For root-canal work the use of the drill can be augmented by the use of a file and, as a bonus, the ultrasonic vibrations produced in the sterilising aqueous irrigation fluid gives a very efficient cleansing action.

1.3.8 Chemical engineering applications

Dispersion of solids

Agglomerates of solid particles can be effectively broken down and dispersed in liquids using power ultrasound. A recent example of this is in the preparation of plastic resins. Scott Bader have used ultrasonic dispersion to achieve a 40% reduction in the amount of an expensive ingredient — pyrogenic silica gel — required to obtain the correct thixotropic behaviour in polyester/polystyrene resins. The ultrasonic homogeniser operates at 12,000 litres per hour.

Filtration

Filtration can be greatly aided by ultrasound. Sonication in the region of the filter itself reduces the gradual clogging which is a common problem and also increases filtration rates.

Crystallisation

When applied to a supersaturated solution, ultrasound has the effect of producing smaller and more uniform crystals than would be formed under conventional conditions. The probable reason for this is the greatly increased nucleation rate induced by cavitation and the formation of large numbers of seeds as newly formed agglomerates are sonically broken down.

Degassing

The degassing of liquids is achieved rapidly upon application of ultrasound. This has potential applications in the beer and soft drink industries. In the laboratory immersion of an open bottle of hplc solvent in an ultrasonic cleaning bath is an excellent way of degassing it before use.

1.4   ULTRASOUND IN CHEMISTRY

Perhaps somewhat surprisingly sonochemistry is not a particularly new subject — it was under active investigation over 50 years ago! There are literature references to applications in polymer and chemical processes in the 1940s [9,10]. Although a book dealing with sonochemistry was translated from Russian and published in 1964 [11], the major renaissance in the subject has occurred only over the last few years. This is undoubtedly due to the more general availability of commercial ultrasonic equipment nowadays. In the 1960's the ultrasonic cleaning bath began to make its appearance in metallurgy and chemical laboratories. Having seen the way in which these baths cleaned soiled glassware and dispersed immiscible organic solvents in aqueous detergent it was not surprising that chemists began to consider using them to enhance chemical reactivity — as indeed we ourselves did in the early 1970s. Over the last few years a large number of articles have been published which describe the variety of applications of power ultrasound in chemical processes [12–19]. These include synthesis (organic, organometallic and inorganic), polymer chemistry (degradation, initiation, and copolymerisation) and some aspects of catalysis. As a result of these revelations the majority of chemists will be aware of the developing science of sonochemistry to the extent that some have already been tempted to experiment with ultrasound in their own laboratories.

One of the first sonochemically assisted syntheses was that reported by Fry in 1978 which involved the use of ultrasonically dispersed mercury in acetic acid for the reduction of α,α'-dibromoketones to α-acetoxyketones [20]. The reaction was performed by dissolving the dibromoketone in acetic acid and dispersing a small amount of mercury in the medium by means of an ultrasonic laboratory cleaning bath for 1–4 days. After this publication, and others of similar type, interest started growing. It was also in the 1970's that biology and biochemistry laboratories began to use ultrasonic cell disruptors on a regular basis. Such instruments offered the chance of introducing greatly increased ultrasonic power into chemical reactions at a modest cost.

For the majority of synthetic chemists interest in sonochemistry will be in power ultrasound since this provides a form of energy for the modification of chemical reactivity which is different from that normally used, for example, heat, light and pressure. Power ultrasound produces its effects via cavitation bubbles. These bubbles are generated during the rarefaction cycle of the wave when the liquid structure is literally torn apart to form tiny voids which collapse in the compression cycle. It has been calculated that pressures of hundreds of atmospheres and temperatures of thousands of degrees are generated on collapse of these bubbles [21]. The physical background for the generation and collapse of such bubbles will be discussed in detail in Chapter 2.

The synthetic chemist will be mainly concerned with reactions in solution and the effects of ultrasound in such cases are best summarised in terms of four different reaction types.

Reactions involving metal surfaces

There are two types of reaction involving metals (i) in which the metal is a reagent and is consumed in the process and (ii) in which the metal functions as a catalyst. While it is certainly true that any cleansing of metallic surfaces will enhance their chemical reactivity, in many cases it has been shown that this effect alone is not sufficient to explain the extent of the sonochemically enhanced reactivity. In such cases it is thought that sonication serves to sweep reactive intermediates, or products, clear of the metal surface and thus present renewed clean surfaces for reaction.

Reactions involving powders or other particulate matter

Just as with the metal surface reactions described above, the efficiency of heterogeneous reactions involving solids dispersed in liquids will depend upon the available reactive surface area and mass transfer.

Until recently, only agitating and stirring with rotating devices and baffled pipes were available techniques to the processors of fluids when mixing, reacting or dissolving small and submicron sized particles on an industrial scale. This can take many hours, days or even weeks until the desired properties are obtained.

The problem with conventional rotational mixing techniques, when trying to disperse solid particles of 10 micrometres in diameter or smaller in a liquid is that the rate of mixing and mass transfer of these particles through the medium reaches a maximum. In fact the mass transfer coefficient k reaches a constant value of about 0.015 cm s^{-1} in water, and liquids of similar viscosity, and cannot be increased any further by increasing the speed of rotational agitation, no matter how large this increase may be. Sonication provides a solution to this problem in that power ultrasound will give greatly enhanced mixing.

A secondary benefit also comes from ultrasonic processing of powders (metallic or non-metallic) — ultrasonic 'pitting' (see Chapter 3) will lead to fragmentation and consequent particle size reduction.

Emulsion reactions

Ultrasound is known to generate extremely fine emulsions from mixtures of immiscible liquids. Ultrasonic homogenisation has been used for many years in the food industry for the production of tomato sauce, mayonnaise and other similarly blended items. In chemistry such extremely fine emulsions provide enormous interfacial contact areas between immiscible liquids and thus the potential for greater reaction between the phases.

Homogeneous reactions

From the above one might be tempted to attribute ultrasonically enhanced chemical reactivity mainly to the mechanical effects of sonication. However this cannot be the whole reason for the effect of ultrasound on reactivity because there are a variety of homogeneous reactions which are also affected by ultrasonic irradiation. How, for example, can we explain the way in which power ultrasound can cause the emission of light from sonicated water (sonoluminescence), the fragmentation of liquid alkanes, the liberation of iodine from aqueous potassium iodide or the acceleration of homogeneous solvolysis reactions?

The answer to these questions lies in the actual process of cavitational collapse. The microbubble is not enclosing a vacuum — it contains vapour from the solvent and any volatile reagents so that, on collapse, these vapours are subjected to the enormous increases in both temperature and pressure referred to above. Under such extremes the solvent and/or reagent suffers fragmentation to generate reactive species of the radical or carbene type some of which would be high enough in energy to fluoresce. In addition the shock wave produced by bubble collapse or even by the propagating ultrasonic wave itself, could act to disrupt solvent structure which could influence reactivity by altering solvation of the reactive species present.

The practising chemist might thus expect to use ultrasound for a range of applications and perhaps achieve one or more of a number of the beneficial effects which are listed in Table 1.2.

Table 1.2 — Possible benefits from the use of power ultrasound in chemistry

- A reaction may be accelerated or less forcing conditions may be required if sonication is applied.
- Induction periods are often significantly reduced as are the exotherms normally associated with such reactions.
- Sonochemical reactions often make use of cruder reagents than conventional techniques.
- Reactions are often initiated by ultrasound without the need for additives.
- The number of steps which are normally required in a synthetic route can sometimes be reduced.
- In some situations a reaction can be directed to an alternative pathway.

REFERENCES

[1] A. P. Cracknell, *Ultrasonics*, Chapter 6, pp. 92–105, 1980, Wykenham Publishers.
[2] J. Curie and P. Curie, *Compt. Rend.*, 1880, **91**, 294; idem., ibid. 1881, 93, 1137.
[3] F. Galton, *Inquiries into Human Faculty and Development*, 1883, Macmillan.
[4] J. Thorneycroft and S. W. Barnaby, *Inst. C. E.*, 1895, **122**, 51.
[5] J. R. Frederick, *Ultrasonic Engineering*, 1965, John Wiley.
[6] F. F. Rawson, *Phys. Bull.*, 1987, **38**, 255.
[7] M. Rule, *The Mary Rose*, 1982, Conway Maritime Press.
[8] J. C. Birnholtz and E. E. Farrell, *American Scientist*, 1984, **72**, 608.
[9] A. Weissler, *J. Chem. Educ.*, 1948, 28.
[10] H. Mark, *J. Acoust. Soc. Amer.*, 1945, **16**, 183.
[11] I. E. El'Piner, *Ultrasound Physical, Chemical and Biological Effects*, 1964, Consultants Bureau, New York.
[12] J.-L. Luche, *L'actualité chimique*, 1982, 21.
[13] P. Boudjouk, *Nachr. Chem. Tech. Lab.*, 1983, **31**, 78.

[14] T. J. Mason, *Ultrasonics*, 1986, **24**, 245.
[15] K. S. Suslick, *Modern Synthetic Methods*, 1986, **4**, 1.
[16] D. Bremner, *Chem. Br.*, 1986, **22**, 633.
[17] R. S. C. Sonochemistry Symposium, University of Warwick, 1986; *Ultrasonics*, 1987, **25**, January.
[18] J. P. Lorimer and T. J. Mason, *J. Chem. Soc., Chem. Soc. Revs*, 1987, **16**, 239.
[19] J. Lindley and T. J. Mason, *J. Chem. Soc., Chem. Soc. Revs*, 1987, **16**, 275.
[20] A. J. Fry and D. Herr, *Tetrahedron Lett.*, 1978, **19**, 1721.
[21] M. A. Margulis, *Russ, J Phys. Chem.*, 1976, **50**, 1.

2

General principles

2.1 INTRODUCTION

In this chapter we will deal with those parts of acoustic wave theory which are relevant to chemists in the understanding of how they may best apply ultrasound to their reaction system. Such discussions will of necessity involve the use of mathematical concepts to support the qualitative arguments. Wherever possible the rigour necessary for the derivation of the basic mathematical equations has been kept to a minimum within the text. For more detailed information the reader is recommended either to read the appendices at the end of the chapter or to consult the references quoted. A summary of the major physical factors influencing sonochemical events has also been included at the end of the chapter for those more interested in the practical applications than the theoretical considerations of ultrasound. Since the vast majority of chemical systems, whether they are homogeneous or heterogeneous, are studied in the solution phase, the discussions here will be restricted to liquid systems.

Being a sound wave, ultrasound is transmitted through any substance, solid, liquid or gas, which possesses elastic properties. The movement of the vibrating body (i.e. sound source) is communicated to the molecules of the medium, each of which transmits the motion to an adjoining molecule before returning to approximately its original position. For liquids and gases, particle oscillation takes place in the direction of the wave and produces *longitudinal* waves (Fig. 2.1(a)). Solids, how-

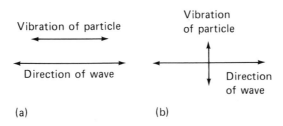

Fig. 2.1 — Wave and particle movement. (a) Transverse waves; (b) longitudinal waves.

ever, since they also possess shear elasticity, can also support tangential stresses giving rise to *transverse* waves, in which particle movement takes place perpendicular to the direction of the wave. (Fig. 2.1(b)).

An easily visualised example of a transverse wave is that obtained when a stone is dropped into a pool of water. The disturbance, or water wave, can be seen spreading across the surface in the form of circular crests of increasing radius. Any objects in the pool (e.g. cork or wood) move up and down when the wave reaches them but they do not move forward in the direction of the wave. In other words, if the motion of particles was considered to be equivalent to the motion of the cork or wood, the particles would move up and down in a direction perpendicular to the horizontal movement of the wave.

A good example of a longitudinal wave can be seen when a coiled spring, anchored at one end, is given a sharp push from the other end. The action causes a disturbance in the spring (Fig. 2.2) which can be seen to 'run' through the whole length.

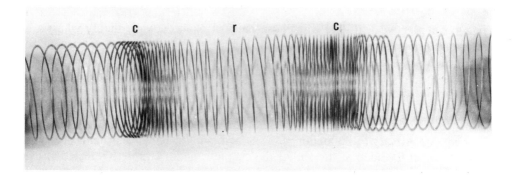

Fig. 2.2 — c = compression; r = rarefaction.

If an individual coil is identified (e.g. by painting it white), as the wave passes the coil it will be seen first to move forward in the direction of the wave and subsequently to return to its original position. For a series of consecutive waves, the motion will be one of oscillation.

The physical aspect of this motion is best understood by examining the action of a tuning fork vibrating in air (Fig. 2.3). Movement of the prong, R, to the right causes the air layer closest to it to be displaced to the right, the disturbed layer pushing the next layer to it, etc. (Fig. 2.3(a)) — i.e. the layers are compressed.

Movement of prong R to the left (Fig. 2.3(c)) causes displacement of the air to the left and hence there is a deficiency of layers to the right of the fork. This is the rarefaction region. The return of R to the right will now begin a new cycle and the wave will proceed with a series of compression and rarefaction portions. It is important to note that the disturbance does not cause the layer to move bodily but to vibrate about a mean rest position — just like the coil of the spring.

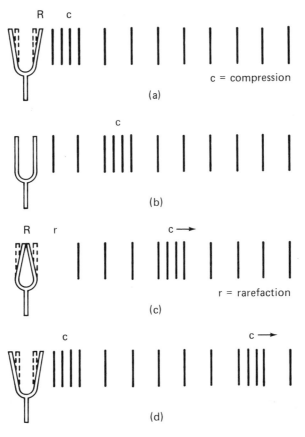

c = compression

(a)

c

(b)

r = rarefaction

(c)

(d)

Fig. 2.3 — Longitudinal waves in air. (a) First outstroke of prongs. (b) Prongs in normal position. (c) First instroke of prongs. (d) Second outstroke of prongs.

At any time (t) the displacement (x) of an individual air molecule from its mean rest position is given by

$$x = x_0 \sin 2\pi ft \tag{2.1}$$

where x_0 is the displacement amplitude, or maximum displacement of the particle, and f is the frequency of the sound wave (Fig. 2.4).

Differentiation of the above leads to an expression for the particle velocity

$$V = dx/dt = v_0 \cos 2\pi ft \tag{2.2}$$

where v_0 ($= 2\pi f x_0$) is the maximum velocity of the particle.

Besides the variation in the molecules' position when the sound wave travels through the air, there is a variation in pressure (Fig. 2.5). At the point where the

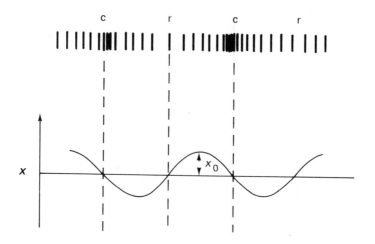

Fig. 2.4 — Displacement (x) graph.

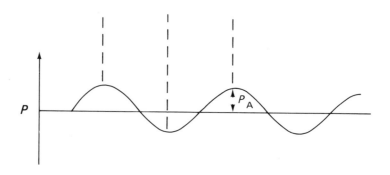

Fig. 2.5 — Pressure (P) graph.

layers are crowded together (i.e. where the molecules are compressed) the pressure is higher than normal at that instant, whereas at the region where the layers are furthest apart (i.e. the rarefaction region) the pressure is lower than normal.

As with displacement, the pressure (P) at any instant is time (t) and frequency (f) dependent:

$$P_a = P_A \sin 2\pi ft \tag{2.3}$$

where P_A is the pressure amplitude.

From Figs 2.4 and 2.5, the maximum particle displacement appears at the point of minimum pressure ($P = 0$) — i.e. the displacement and pressure are out of phase. Although this may at first cause some confusion it may be easily demonstrated by reference to Fig 2.6.

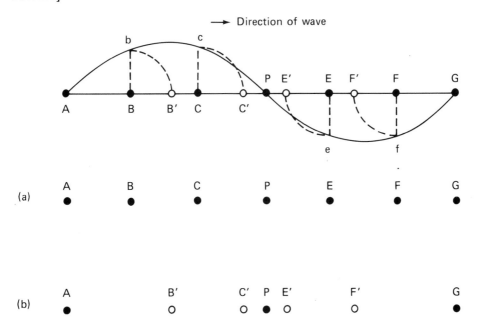

Fig. 2.6 — Displacement of a longitudinal wave. ●, Original particle position. ○, Displaced particle position.

On passage of the wave, the particles, originally at the rest positions A, B, C etc. are displaced to their new positions A, B′, C′ etc. Displacements such as BB′ (or CC′) which are to the right and in the direction of the wave are represented by lines such as Bb and Cc above the x-axis. Displacements such as EE′, to the left and in the opposite direction to the wave's movement are drawn as Ee below the x-axis — i.e. negative displacements. In the region P the particles (C′, P, E′) are crowded together and there is compression — i.e. high pressure values. At A the particles are more separated than normal and there is rarefaction, i.e. decreased pressure. The displacement and pressure are out of phase.

Perhaps the most easily visualised demonstration of the out-of-phase nature of displacement and pressure is obtained by attaching a supported weight on a spring (Fig. 2.7(a)).

On removing the support the spring will immediately elongate due to the force (equivalent to pressure) of the weight acting on it. At maximum extension the overall force (or pressure) acting on the spring is zero (Fig. 2.7(b)). On recoil the spring contracts, lifting the weight upwards, and the force increases in magnitude but in the negative sense. At the instant of maximum contraction (i.e. negative displacement) the overall force on the spring is again zero (Fig. 2.7(d)).

One of the most important characteristics necessary to completely identify a wave is its intensity, where the intensity is a measure of the sound energy the wave

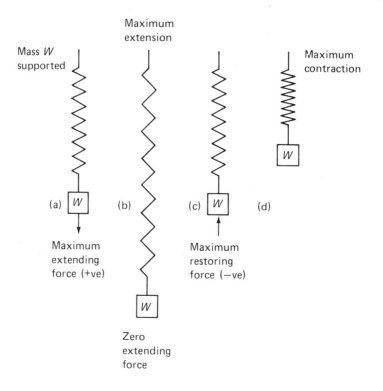

Fig. 2.7

produces. For a sound wave in air, the mass (m) of air moving with an average velocity (v) will have associated with it a kinetic energy of (mv^2)/2 (joules). In the strictest sense the intensity is the amount of energy carried per second per unit area by the wave. Since the units of energy are joules (J) and a joule per second is a watt (W), then the usual unit of sound intensity (especially in sonochemistry) will be $W\,cm^{-2}$. As we will see later, equation (2.13), the maximum intensity (I) of the sound wave is proportional to the square of the amplitude of vibration of the wave (P_A^2). This will have important repercussions in our study of chemical systems.

Let us now turn our attention to the application of the sound wave to a liquid since this is the medium of importance to the practising chemist. The sound wave is usually introduced to the medium by either an ultrasonic bath or an ultrasonic horn (see Chapter 7). In either case, an alternating electrical field (generally in the range 20–50 kHz) produces a mechanical vibration in a transducer, which in turn causes vibration of the probe (or bottom of the bath) at the applied electric field frequency. The horn (or bath bottom) then acts in a similar manner to one prong of a tuning fork.

As in the case of air, the molecules of the liquid, under the action of the applied acoustic field, will vibrate about their mean position and an acoustic pressure ($P_a = P_A \sin 2\pi ft$) will be superimposed upon the already ambient pressure (usually hydrostatic, P_h) present in the liquid. The total pressure, P, in the liquid at any time, t, is given by:

$$P = P_h + P_a \tag{2.4}$$

where P_a is the applied acoustic pressure (eqn. 2.3). The displacement (x) and the velocity (v) of the particles are given, as before, by equations (2.1) and (2.2).

It is worth noting here, that for the ultrasound, as for any sound wave, the wavelength of sound in the medium is given by the relationship

$$c = \lambda f \tag{2.5}$$

For the frequencies usually employed to influence chemical processes (20–50 kHz), the wavelengths produced in the liquid medium are in the range 7.5–3.0 cm. For diagnostic or relaxation investigations (see later), with greater frequencies (1–100 MHz), the wavelength range will be 0.15–0.0015 cm. These wavelengths are considerably longer than bond length values and sonochemical effects are not therefore the result of direct interactions between the reagent and the wave as is the case in photochemistry.

2.2 INTENSITY AND PRESSURE AMPLITUDE

In our previous discussions (tuning fork in air) we pointed out that sound was a form of energy. The particles of the medium were set into vibratory motion and thereby possessed kinetic energy. This energy was derived from the wave itself. Using this principle we can deduce the energy (and hence intensity) associated with our applied ultrasonic field.

Consider the movement of a layer of the medium of area A and thickness dx, (i.e. volume $A dx$) under the action of the ultrasonic wave (Fig. 2.8).

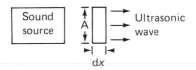

Fig. 2.8

Then the kinetic energy $(mv^2/2)$ of the layer is given by:

$$KE = 1/2(\rho A dx)v^2 \tag{2.6}$$

The energy for the whole wave E_t may be obtained by summing all such elements (i.e. integrating equation (2.6)) to give

$$E_t = 1/2\rho A x v^2 \tag{2.7}$$

and the energy per unit volume (Ax) or energy density, E, given by

$$E = 1/2\rho v^2 \tag{2.8}$$

If the sound energy passes through unit cross-sectional area $(A = 1)$ with a velocity of c, then the volume swept out in unit time is c (since $A = 1$), and the energy flowing in unit time is given by Ec. Since intensity (I) has been defined as the amount of energy flowing per unit area $(A = 1)$ per unit time, then

$$I = Ec \tag{2.9}$$

and from equation (2.8)

$$I = 1/2\rho c v^2 \tag{2.10}$$

For a plane progressive wave, the particle velocity, v, can be shown [1] to be related to the acoustic pressure, P_a, by the expression

$$P_a/v = \rho c \tag{2.11}$$

where p is the density of the medium and c the velocity of sound in the medium. (The derivation of equation (2.11) is given in Appendix 1.) For maximum particle velocity, v_0, the amplitude of the oscillating acoustic pressure, P_A, is given by

$$P_A/v_0 = \rho c \tag{2.12}$$
i.e.
$$v_0 = P_A/\rho c \tag{2.12a}$$

Thus the intensity of the sound wave (from eqns (2.10) and (2.12a)), may be expressed as

$$I = P_A^2/2\rho c \tag{2.13}$$

i.e. the sound intensity is proportional to the square of the acoustic amplitude.

Clearly, to measure the sound intensity at a particular point in a medium, either the maximum particle velocity, v_0 (eqn. 2.2) or the maximum pressure amplitude, P_A (eqn. 2.13) must be determined. In practice, this is extremely difficult and for most sonochemical applications a calorimetric determination of the total ultrasonic energy delivered to the medium is considered to be sufficient (see later).

As an example of the use of equation (2.13) let us consider the passage of a wave, of frequency 20 kHz and intensity 1 W cm^{-2}, through water at room temperature. If we take the density of water, ρ, to be 1000 k gm^{-3} and the velocity of sound, c, as 1500 m s^{-1}, then the maximum pressure amplitude, P_A ($= (2\rho cI)^{1/2}$) will be 1.73×10^{-5} N m^{-2}. This means the acoustic pressure varies from approximately $+1.7$ atmospheres to -1.7 atmospheres twenty thousand times per second. The maximum particle velocity, v_0 (eqn. 2.12) and displacement amplitude, x_0, ($v_0 = 2\pi f x_0$) can be calculated to be 11.7 m s^{-1} and 9.31×10^{-5} cm respectively. Also it can easily be shown that the particle acceleration, a ($= dv/dt$) has an acceleration amplitude (a_0) given by $a_0 = 4\pi^2 f^2 x_0$. For the above values of f, ρ and c, this yields a value of 1.58×10^4 m s^{-2}, an acceleration which is approximately 1600 times greater than under the action of gravity.

2.3 SOUND ABSORPTION

During the propagation of a plane sound wave through a medium the intensity of the wave decreases as the distance from the radiation source increases. The intensity, I, at some distance, d, from the source is given by:

$$I = I_0 \exp(-2\alpha d) \tag{2.14}$$

where α is the absorption (attenuation) coefficient. This attenuation may arise as a result of reflection, refraction, diffraction or scattering of the wave or it may be the result of converting some of the mechanical (kinetic) energy of the wave into heat. For chemical applications, which usually take place in the gaseous or liquid phase, it is the latter process which is the most important. As the molecules of the medium vibrate under the action of the sound wave, they experience viscous interactions which degrade the acoustic energy into heat, and it is the absorption of this degraded acoustic energy by the medium which gives rise to the small observed bulk heating effect during the application of high power ultrasound. In practice the experimental temperature often rises very quickly (approximately 5°C) during the first few minutes of applying ultrasound. (After this initial period the temperature remains effectively constant, provided the reaction vessel is effectively thermostatted.) According to Stokes [2], the absorption coefficient in a liquid due to frictional losses, α_s, is given by

$$\alpha_s = 8\eta_s \pi^2 f^2 / 3\rho c^3 \tag{2.15}$$

where η_s is the ordinary (or shear) viscosity of the liquid.

Kirchhoff [3] has suggested that energy losses due to heat (thermal) conduction in the medium must also be considered. At any instant the high pressure region will have a temperature above the average while the temperature of the low pressure regions will be below average. Heat will therefore be conducted from the high to low temperature regions and a compressed region will return less work on expansion than was required to compress it. This leads to a sound absorption coefficient, $\alpha_{th} = 2\pi^2 K(\gamma - 1)f^2/(\rho\gamma C_v c^3)$.

The total loss, or absorption, caused by both viscosity and thermal conductivity is called the classical absorption, α_{cl}, and is given by

$$\alpha_{cl} = \alpha_s + \alpha_{th} = \frac{2\pi^2}{\rho c^3}\left\{\frac{4}{3}\eta_s + \frac{(\gamma-1)K}{\gamma C_v}\right\}f^2$$

Since $C_p/C_v = \gamma$ — i.e. $\gamma C_v = C_p$ then the above may be written as:

$$\alpha_{cl} = \frac{2\pi^2 f^2}{\rho c^3}\left\{\frac{4}{3}\eta_s + \frac{(\gamma-1)K}{C_p}\right\} \tag{2.16}$$

However when values of the calculated absorption coefficient are compared with those obtained experimentally, the agreement is often poor. For example if we take water at 20°C for which $\eta_s = 1\ cp$, $\rho = 1\ \mathrm{g\,cm^{-3}}$ and $c = 1500\ \mathrm{m\,s^{-1}}$, and we pass a sound wave of 20 kHz, then α can be calculated to be approximately 3.5×10^{-8} cm^{-1}. Experimentally α is found to be 8.6×10^{-8} cm^{-1} — i.e. approximately two-and-a-half times larger. In fact only in the case of monatomic gases is the observed absorption, α_{obs}, equal to the classical absorption. In all other cases the observed absorption is greater than the classical absorption by an amount called the excess absorption, α_{ex} ($= 2\pi^2 f^2 \eta_B/\rho c^3$). For complete accuracy, equation (2.16) should be further modified to take account of the compressional viscous forces which act during rarefaction and compression.

That is

$$\alpha = \frac{2\pi^2 f^2}{\rho c^3}\left\{\frac{4}{3}\eta_s + \eta_B + \frac{(\gamma-1)K}{C_p}\right\} \tag{2.17}$$

where η_B is termed the bulk viscosity.

The bulk viscosity here, η_B, should not be confused with the so-called bulk viscosity of polymers which refers to the steady flow shear viscosity of the bulk undiluted polymer. Here it represents all the causes of sound absorption other than those by shear viscosity or thermal conductivity. Typically these may be:

(1) Energy losses associated with the flow of liquid molecules between positions of different density.
(2) Relaxation processes, e.g. rotational isomerisation or vibrational energy transfer. Relaxation will be dealt with later.

According to the above expressions the value of α/f^2 is a constant for a given liquid at a given temperature. Any increase in sound frequency, f, must result in a compensatory increase in α and thus a more rapid attenuation of the sound intensity with distance (eqn. 2.14). This has important consequences. Consider for example the passage of sound through water at room temperature. According to Fox and Rock [4] the value of α/f^2 for a wide variety of frequencies in water, is 21.5×10^{-17} cm^{-1}. Using this value the absorption coefficients at 21.5 kHz and 127.0 kHz can be deduced to be 9.9×10^{-8} cm^{-1} and 3.47×10^{-6} cm^{-1} respectively, and the penetration depths (eqn. 2.14) at which the sound intensities are reduced to half of their original values to be 35 km and 1 km respectively. Calculations such as these demonstrate clearly that in order to achieve identical intensities at a given depth (distance) in a liquid, it will be necessary to use a higher initial power for the source with the higher sound frequency. For example, to achieve an intensity of 20 W cm^{-2} at a liquid depth of 10 cm (typical sonochemical reaction vessel) using sound sources of 10 kHz and 10 MHz, will require initial intensities of 20 and 30.7 W cm^{-2} respectively. Moving to higher frequencies still (e.g. 20 MHz) requires even higher initial intensities (e.g. approximately 112 W cm^{-2}). However, although the constancy of α/f^2 is satisfied for many liquids, some highly structured liquids absorb more energy at one particular frequency than another and the value of α/f^2 is found to vary with the applied frequency, f. This variation in α/f^2 with f has been used to investigate the structure of the liquids and we will return to this topic later in Chapter 6, which deals with relaxation phenomena.

2.4 BUBBLE FORMATION AND THE FACTORS AFFECTING CAVITATION THRESHOLD

2.4.1 Effect of gas and particulate matter

It was suggested previously that the progression of a sound wave through a liquid medium caused the molecules to oscillate about their mean position. During the compression cycle, the average distance between the molecules decreased, whilst during rarefaction the distances increased. If a sufficiently large negative pressure, P_c, is applied to the liquid (here it will be the acoustic pressure on rarefaction, $P_c = P_h - P_a$), such that the average distance between the molecules exceeds the critical molecular distance (R) necessary to hold the liquid intact, the liquid will break down and voids or cavities will be created — i.e. cavitation bubbles will be formed. The production of such bubbles has been known for many years and a good example is provided by either a ship's propeller or a paddle stirrer where the cavities are produced by the rapid rotation of the blade through the liquid. Once produced, these cavities, voids, or bubbles, may grow in size until the maximum of the negative pressure has been reached. In the succeeding compression cycle of the wave however, they will be forced to contract, i.e. decrease in volume and some of them may even disappear totally. (In other cases bubble oscillations may result. See later.) The shock waves produced on total collapse of the bubbles has been estimated to be of the order of several of thousands of atmospheres and is thought to be the cause of the considerable erosion observed for components in the vicinity of the bubble. (Fig. 2.9 shows the erosion of a 1 cm diameter titanium alloy probe used in our laboratories after running at 20 W cm^{-2} for 20 h in benzene.)

Estimates of the acoustic pressure necessary to cause cavitation in water (see Appendix 2) has led to a value of approximately 1500 atm. In practice, cavitation occurs at considerably lower values (< 20 atm) and this is undoubtedly due to the presence of weak-spots in the liquid which lower the liquid's tensile strength. There is

Fig. 2.9 — Erosion of 1 cm diameter titanium alloy tip. (a) Initial surface. (b) Eroded surface
(20 hours at 20 W cm^{-2} in benzene).

now sufficient experimental evidence to suggest that one cause of weak-spots is the presence of gas molecules in the liquid. For example, it has been observed that the degassing of liquids has led to increases in the cavitation threshold — i.e. to increases in the values of the applied acoustic pressure necessary before cavitation bubbles were observed. Further, the application of external pressures which would cause any suspended gas molecules to dissolve, thereby effectively removing the gas nuclei, has also been found to lead to increases in the cavitation threshold. It can be argued for the pressure experiments that since cavitation can only be produced when the (negative) acoustic pressure exceeds the 'liquid' pressure holding the liquid intact, the application of an external pressure (i.e. pressurising the system) will necessitate the application of a higher negative acoustic pressure (P_a). This in turn will require the use of higher intensity sound waves ($P_a = P_A \sin 2\pi ft : P_A^2 = 2Ipc$), to overcome the liquids' cohesive forces — i.e. the threshold is seen to be raised.

It has also been found that the presence of particulate matter, and more especially the occurrence of trapped vapour-gas nuclei in the crevices and recesses of these particles, also lowers the cavitation threshold. The way in which nucleation occurs at these sites (and from similar sites on the vessel walls) is shown in Fig. 2.10.

During the rarefaction cycle of the acoustic wave, as the pressure in the liquid decreases the liquid gas interface becomes increasingly more convex, its angle of contact decreases, until, at sufficiently low pressure it breaks away from the surface to produce a bubble of radius, R_i.

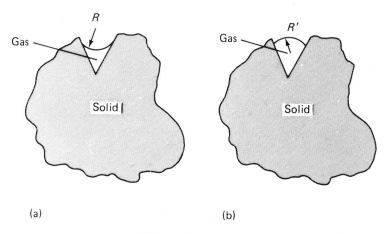

Fig. 2.10 — Crevice model for stabilising cavitation nuclei. (a) For external positive pressure. (b) For external negative pressure.

Attempts to totally remove all particulate matter (i.e. ultrafiltration) have not been completely successful in that the theoretical limit for water's tensile strength (~ 1500 atm) has not been achieved. The largest experimental threshold value (~ 200 atm) is that found by Greenspan [5].

In the absence of all potential nuclei for cavitation (i.e. gas or particulate matter) it is possible to deduce an equation relating the critical pressure (P_K) which must be exceeded to create a bubble (or void) of radius R_e in an ultrapure liquid. Provided vapour pressure is neglected then:

$$P_K = -\frac{2}{3}\left\{\frac{(2\sigma/R_e)^3}{3(P_h + 2\sigma/R_e)}\right\}^{1/2} \tag{2.18}$$

(The fact that the pressure is negative implies that a negative pressure must be applied to overcome the cohesive forces of a liquid. Those readers interested in the derivation of this equation should turn to Appendix 2.)

To create a bubble in water ($\sigma = 0.076\ \text{N}\,\text{m}^{-1}$) of radius 10^{-4} cm will require a rarefaction pressure (i.e. negative pressure) in excess of 1.78 atm. Provided the maximum rarefaction pressure (P_A) is larger than 1.78 atm, the bubble created will expand during the rarefaction cycle. During this expansion cycle, liquid vapour may evaporate into the partial void. Thus there will probably be several different types of cavity in the liquid:

(a) the empty cavity (true cavitation),
(b) the vapour-filled cavity,
(c) the gas-filled cavity, unless the liquid is totally degassed,
(d) a combination of vapour and gas filled cavities.

It is the subsequent fate of some of these bubbles, as they oscillate in the applied sinusoidal acoustic field, which is the origin of sonochemistry. However before embarking upon a discussion of bubble dynamics let us consider what other factors apart from degassing, pressurising and filtration affect the onset of cavitation.

2.4.2 Effect of viscosity

Since it is necessary for the negative pressure in the rarefaction cycle to overcome the natural cohesive forces acting in the liquid, any increase in these forces will increase the threshold of cavitation. One method of increasing these forces is to increase the viscosity of the liquid. Table 2.1 below shows the influence of viscosity on the

Table 2.1 — Sound pressure (P) producing cavitation in various liquids under a hydrostatic pressure of 1 atm.

Liquid	η (poise)	ρ (g/cm^{-3})	c (km/s^{-1})	P_A (atm)
Castor oil	6.3	0.969	1.477	3.9
Olive oil	0.84	0.912	1.431	3.61
Corn oil	0.63	0.914	1.463	3.05
Linseed oil	0.38	0.921	1.468	2.36
CCl$_4$	0.01	1.60	0.926	1.75

pressure amplitude (P_A) at which cavitation begins in several liquids at 25°C, at a hydrostatic pressure of 1 atmosphere.

The effect, though not insignificant, is hardly dramatic. Taking corn and castor oils as examples, a tenfold increase in viscosity has led only to a 30% increase in the acoustic pressure needed to bring about cavitation.

2.4.3 Effect of applied frequency

To completely rupture a liquid and hence provide a void, which may subsequently become filled with gas or vapour, requires a finite time. For sound waves with high frequencies, the time required to create the bubble may be longer than that available during the rarefaction cycle. (For example, at 20 kHz the rarefaction cycle lasts 25 μs ($= 1/2f$), attaining its maximum negative pressure in 12.5 μs, whereas at 20 MHz the rarefaction cycle lasts only 0.025 μs.) Thus it might be anticipated that as the frequency increases the production of cavitation bubbles becomes more difficult to achieve in the available time and that greater sound intensities (i.e. greater amplitudes) will need to be employed (over these shorter periods), to ensure that the cohesive forces of the liquid are overcome.

Fig. 2.11, where the variation in threshold intensity with frequency is shown for both aerated and gas free water, demonstrates this quite clearly.

As expected the threshold for aerated water is lower than that for gas-free water (see earlier) and the threshold intensity increases with increase in frequency. In fact ten times more power is required to make water cavitate at 400 kHz than at 10 kHz. It

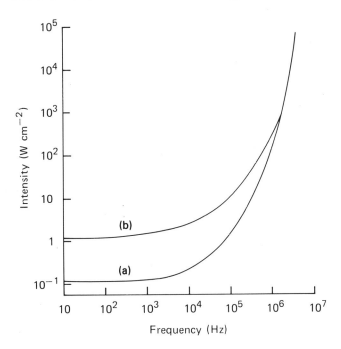

Fig. 2.11 — Variation in threshold intensity with frequency. (a) Aerated water. (b) Air-free water.

is for this reason that 20–50 kHz frequencies were generally chosen for cleaning purposes and have subsequently been found of value in sonochemistry. As can be seen (Fig. 2.11) there is little extra energy needed to cause water to cavitate at 50 kHz than to cavitate at 10 kHz. However, below 16 kHz there may be some noise discomfort to the user. The ear responds to mechanical vibration in the range 16–16 000 Hz, and hence frequencies greater than 16 kHz are usually employed in sonochemical applications.

Most chemists working in the field of sonochemistry, if not employing an ultrasonic bath, make use of a commercial probe system. These instruments often possess a pulse facility and are specifically designed for biological cell disruption where temperature control is important. This pulse facility enables the power ultrasound to be delivered intermittently and thereby allow periods of cooling. The time (i.e. pulse length) for which the sound energy is delivered to the system is controlled by an instrument setting and may be varied from 0%, where no energy is supplied, to 100%, which is continuous application of the energy. (See Chapter 7.) For sonochemical applications, however, there is a minimum time period (pulse) which must be exceeded if any cavitational effects are to be observed. This is due to the time delay between the application of the acoustic excitation (i.e. the sound wave) and the onset of cavitation, and the pulse of acoustic energy may not be present long enough to create the cavitation bubble. This is best visualised by considering Fig. 2.12 which illustrates the growth of a bubble of radius 8×10^{-5} cm in

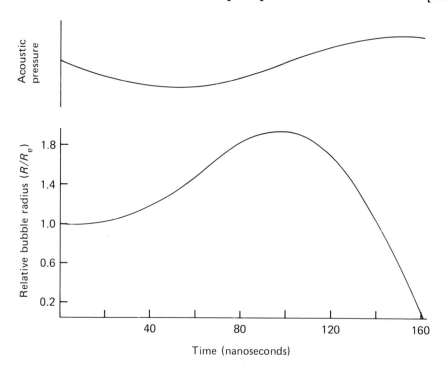

Fig. 2.12 — Radius–time curve for an air bubble ($R_e = 8 \times 10^{-5}$ cm) sonicated in water at 5 MHz and 5.5 W cm^{-2}. $P_A = 4$ atm; $f = 5$ MHz; $R_e = 8 \times 10^{-5}$ cm.

water at an ambient pressure of 1 atmosphere and subject to an applied acoustic field of frequency 5 MHz and a pressure amplitude (P_A) of 4 atm.

For the first eighth of the acoustic cycle (i.e. 25 ns) there is only a small increase in the size of the bubble ($\sim 5\%$). Even after quarter of a cycle (i.e. 50 ns) it has only grown by $\sim 30\%$. During the next quarter of a cycle it grows to a size sufficiently large, approximately twice its initial radius, that under the action of the applied acoustic field it collapses totally — i.e. cavitation takes place. Thus, if the pulse is present for less than quarter a cycle (50 ns) at this frequency, the bubble will not have sufficient time to grow to a size capable of collapse. (More will be said about this later.) In general the threshold intensity is found to decrease as the pulse length is increased and an upper limit for pulse length is usually attained after which the threshold remains independent of the length of the pulse. For 20 kHz, the upper limit is approximately 20 ms.

2.4.4 Effect of temperature
The final factor to be considered here, and known to affect the cavitation threshold, is the temperature. In general, the threshold limit has been found to increase with decrease in temperature. This may in part be due to increases in either the surface tension (σ) or viscosity (η) of the liquid as the temperature decreases, or it may be

due to the decreases in the liquid vapour pressure (P_v). To best understand how these parameters (σ, η, P_v) affect the cavitation threshold, let us consider an isolated bubble, of radius R_0, in water at a hydrostatic pressure (P_h) of 1 atm.

Any bubble within a liquid is subject to both the crushing force of the hydrostatic pressure (P_h) and those due to surface tension effects ($2\sigma/R_0$). In order that the bubble should remain in equilibrium, the supporting forces due to the pressure of gas (P_g) and vapour (P_v) in the bubble must equal the crushing forces — i.e.

$$P_v + P_g \;=\; P_h + 2\sigma/R_0 \qquad\qquad (2.19)$$

Obviously, if the pressure within the bubble ($P_v + P_g$) exceeds those trying to collapse the bubble ($P_h + 2\sigma/R_0$), the bubble will expand (and vice versa). In other words a bubble will grow (expand) when ($P_v + P_g$) is greater than ($P_h + 2\sigma/R_0$)—i.e.

$$P_v + P_g \;>\; P_h + 2\sigma/R_0 \qquad\qquad (2.20)$$

or

$$P_v \;>\; P_h + 2\sigma/R_0 - P_g \qquad\qquad (2.21)$$

If we neglect, momentarily, any surface tension effects ($2\sigma/R_0 \sim 0$) and assume that the liquid contains only a small amount of gas ($P_g \sim 0$), then we may deduce that 'expanding' bubbles are created in a liquid when the vapour pressure exceeds the atmospheric pressure ($P_v > P_h$). For water the vapour pressure at 100°C is 1 atm and hence water, at a hydrostatic pressure of 1 atmosphere, boils as soon as the temperature exceeds 100°C. At 25°C the vapour pressure of water is 0.023 atm and thus water will only boil, at 25°C, if the atmospheric pressure is less than this value. This can readily be achieved by evacuating the system.

Let us now consider the effect of applying an ultrasonic wave to the liquid. The pressure within the liquid will now become ($P_h + P_a$), where P_a ($= P_A \sin 2\pi ft$) is time dependent. During the compression cycle of the wave P_a is positive, rising from zero to P_A in quarter of a cycle before falling to zero after half a cycle. During this time the pressure in the liquid will have risen from P_h to $P_h + P_A$ before returning again to P_h. In the rarefaction cycle P_a will become negative, the pressure in the liquid at any time being given by $P_h - P_a$. (For example, after threequarters of a cycle the wave will be at the maximum of its rarefaction cycle and the liquid pressure will be $P_h - P_A$).

Hence in the presence of an acoustic field, equation (2.21) becomes

$$P_v \;>\; (P_h - P_a) + 2\sigma/R_0 - P_g \qquad\qquad (2.22)$$

or neglecting surface tension effects and gas pressure,

$$P_v \;>\; P_h - P_a \qquad\qquad (2.23)$$

i.e. the liquid boils (produces cavitation bubbles), when the vapour pressure exceeds that of ($P_h - P_a$). For the examples of water at 100°C ($P_v = 1$ atm) and 25°C ($P_v =$

0.023 atm), the magnitudes of the applied acoustic pressure (P_a) will be approximately zero and 1 atmosphere respectively. In other words, a greater intensity, I ($= P_a^2/2\rho c$), will be necessary to cause water to cavitate at the lower temperature.

If surface tension is not neglected equation (2.22) may be represented as

$$(P_v - 2\sigma/R_0) = P'_v > P_h - P_a \tag{2.24}$$

Thus if we employ different liquids with decreasing surface tensions (σ), the value of P'_v (assuming constant P_v) will increase and lower P_a values, at a given temperature, will need to be applied before P'_v exceeds ($P_h - P_a$).

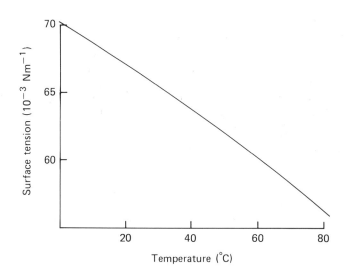

Fig. 2.13 — Variation of the surface tension of water with temperature.

For water the surface tension varies with temperature as shown in Fig. 2.13 — i.e. a lowering of surface tension with increase in temperature. If it can be assumed that P_v remains constant with increase in temperature then there will be a small increase in P'_v and a lowering of the intensity (P_a) necessary to cause cavitation. Obviously P_v does not remain constant with increase in temperature but increases quite rapidly. Consequently there is a rapid rise in P'_v with increase in temperature and the threshold decreases accordingly. The corollary is that liquids with high vapour pressures or low surface tensions cavitate at a lower intensity.

Let us now consider the effect of solvent viscosity on the cavitation threshold. According to Table 2.1, an increase in the solvent viscosity required the application of a greater sound intensity (i.e. P_a) before cavitation bubbles were observed. For example, castor oil ($\eta = 0.63$ N m^{-1}) and corn oil ($\eta = 0.063$ N m^{-1}) require acoustic pressures of 3.9 and 3.05 atm respectively. For water, viscosity decreases with increased temperature, and hence lower P_a values (i.e. lower intensities) will be needed to cause cavitation as the temperature is increased. Overall then, the general conclusion is that cavitation bubbles are more easily produced as the temperature is raised. However the sonochemical effects of such bubbles may be reduced — see Section 2.6.3.

Before leaving this section let us consider the effect of increasing the atmospheric pressure — i.e. pressurising the system. In our previous discussions we suggested that pressurising the system raised the cavitation threshold. A consideration of equation (2.23) shows why this is so. If a liquid will only create cavitation bubbles when $P_v > P_h - P_a$, then any increase in P_h (i.e. pressurising) will by necessity require larger P_a values and hence larger intensities.

Having identified the factors affecting the production of these microbubbles or voids, we must now turn our attention to their fate in the presence of the oscillating sinusoidally applied acoustic field. For any bubble created early in the rarefaction cycle, or initially present in the liquid, a growth in size will occur during the remainder of the cycle. During the compression cycle all bubbles will be made to contract or collapse. However, if during growth gas, or vapour has diffused into the void or bubble, complete collapse may not occur and the bubble may in fact oscillate in the applied field. We must therefore consider two types of bubble; those which collapse completely (TRANSIENT) and those which oscillate and exist for some considerable period of time (STABLE). Whether they collapse or oscillate depends upon many factors, e.g. temperature, acoustic amplitude, frequency, external pressure, bubble size, gas type and content. There is a further complication in that transient bubbles may grow into stable bubbles and vice versa. What is certain however, is that compression of a bubble, containing gas or vapour, occurs very rapidly and leads to enormous temperatures ($\sim 10\,000$ K) and pressures (~ 1000 atm) within the bubble itself, and that on complete collapse, should it occur, these pressures must be released as shock waves into the liquid.

2.5 MOTION OF THE BUBBLE IN THE APPLIED ACOUSTIC FIELD

The important breakthrough in the understanding of cavitation came in 1917 when Lord Rayleigh [6] published his paper 'On the pressure developed in a liquid during the collapse of a spherical cavity'. By considering the total collapse of an empty void under the action of a constant ambient pressure P_0, Rayleigh deduced both the cavity collapse time τ, and the pressure P in the liquid at some distance R from the cavity to be respectively,

$$\tau = 0.915 R_m (\rho/P_0)^{1/2} \tag{2.25}$$

$$P/P_0 = 1 + (R/3r)(Z - 4) - (R^4/3r^4)(Z - 1) \tag{2.26}$$

where R_m is the radius of the cavity at the start of collapse (see Appendix 3), $Z = (R_m/R)^3$ and $r = R/R_m$.

Schematically, equation (2.26) may be represented by Fig. 2.14 with the maximum in pressure, p_{max}, occurring at a distance $4^{1/3}R$ from the cavity. Equation (2.25) is useful in that it allows an estimate of the point in the compression cycle where total

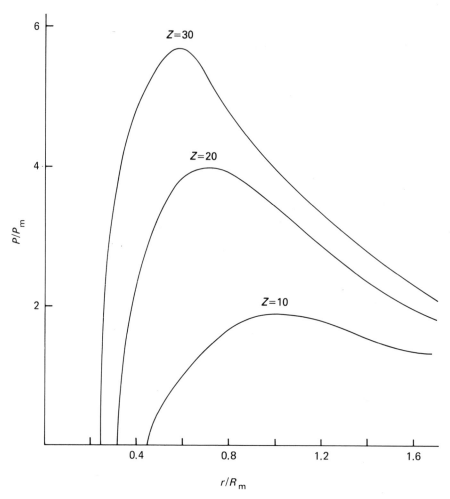

Fig. 2.14 — Pressure developed in a liquid surrounding a collapsing Rayleigh cavity. Z = volume compression ratio.

collapse is likely to occur. For example, the collapse time of a bubble of radius 10^{-3} cm (R_m), in water at an ambient pressure (P_0) of 1 atmosphere, is approximately 1 μs. Since an applied acoustic field of 20 kHz has a compression cycle of 25 μs, it is expected that total collapse will occur in the first 4% of the cycle.

However, this can only be an estimate since it is unlikely that a bubble in a sound field will feel a constant pressure, P_0, exerted during its collapse (acoustic pressure is time dependent — i.e. $P_a = P_A \sin 2\pi f t$) nor will it be an empty void, being filled with either gas or vapour. As such, equation (2.25) must be modified to give:

$$\tau = 0.915 R_m (\rho/P_m)^{1/2} (1 + P_{vg}/P_m) \qquad (2.27)$$

where P_m is the pressure in the liquid and P_{vg} the pressure in the bubble at the start of bubble collapse (see Appendix 3). If it can be assumed that the bubble collapse time is small in comparison with the time period of the compression cycle then P_m can be assumed to be constant and may be approximated to $(P_a + P_h)$. For an empty void (i.e. $P_{vg} = 0$) in the absence of an applied acoustic field (i.e. $P_a = 0$) then P_m may be replaced by P_h ($= P_0$) and equation (2.27) reverts to the Rayleigh equation.

In any cavitation field most of the visible bubbles will be oscillating stably and it is perhaps pertinent that we concentrate our discussions first on the fate of such bubbles in the acoustic field. If we assume that we have a bubble with an equilibrium radius, R_e, existing in a liquid at atmospheric pressure P_h, then the oscillation of the bubble and in particular the motion of the bubble wall, under the influence of the applied sinusoidal acoustic pressure (P_a) is a simple dynamical problem, akin to simple harmonic motion for a spring.

Although there exist many sophisticated mathematical treatise which derive the motion of the bubble wall, all yield equations similar in form to

$$R\ddot{R} + (3/2)\dot{R}^2 = 1/\rho \left[\left(P_h + \frac{2\sigma}{R_e} - P_v \right) \left(\frac{R_e}{R} \right)^{3K} - \frac{2\sigma}{R} - 4\eta\frac{\dot{R}}{R} - (P_h - P_a) \right]$$

$$(2.28)$$

where $\dot{R} = dR/dt$ = velocity of the cavity wall. $\ddot{R}$ is the acceleration of the cavity wall, R_e is the radius of the bubble at its rest (equilibrium) position, σ is the surface tension of the liquid, η is the viscosity of the liquid, P_v is the vapour pressure of the liquid, ρ is the density of the liquid, P_h is the atmospheric (hydrostatic) pressure, P_a is the applied acoustic pressure and K is the polytropic index of the gas. (K varies between γ, the specific heat ratio, and unity, the limit for adiabatic and isothermal conditions.) [For the reader interested in the derivation of this equation, see Appendix 4.]

Neglecting the vapour pressure and viscosity contributions and replacing P_a by $P_A \sin w_a t$ gives

$$R\ddot{R} + (3/2)\dot{R}^2 = 1/\rho \left[\left(P_h + \frac{2\sigma}{R_e} \right) \left(\frac{R_e}{R} \right)^{3K} - \frac{2\sigma}{R} - P_h + 2P_A \sin \omega_a t \right] \quad (2.29)$$

where w_a ($= 2\pi f_a$) is the applied circular frequency.

Provided we neglect damping effects (viscous and thermal), equation (2.29) adequately describes the motion of stable bubbles over several cycles. Before proceeding to discuss the effect that P_h, σ, ρ, P_A, f and R_e, have on the solutions of equation (2.29) (i.e. radius-time variation), it may be instructive to consider here a simple modification of the equation such that with the aid of a computer one might deduce how the variation in some of the above parameters affects the radius-time of the bubble.

Let us assume that the bubble on expansion (or contraction) increases (or decreases) its radius by an amount r, such that R, the bubble radius at any time is given by $R = R_e + r$. Then provided $r \ll R_e$, substitution into equation (2.29), followed by expanding in powers of $1/R_e$ (and retaining only the first order terms) gives:

$$\ddot{r} + w_r^2 r = (P_A/\rho R_e)\sin\omega_a t \qquad (2.30)$$

where w_a is the equal applied circular frequency ($= 2\pi f_a$) and w_r is the resonant frequency of the bubble. For small amplitude vibrations the resonance frequency is given by:

$$\omega_r^2 = \left(\frac{1}{\rho R_e^2}\right)\left[3K\left(P_h + \frac{2\sigma}{R_e} - P_v\right) - \frac{2\sigma}{R_e} - \frac{4\eta^2}{\rho R_e^2}\right] \qquad (2.31)$$

Neglecting the effects of viscosity (η) and solvent vapour pressure (P_v) equation (2.31) reduces to:

$$\omega_r^2 = \frac{1}{\rho R_e^2}\left[3K\left(P_h + \frac{2\sigma}{R_e}\right) - \frac{2\sigma}{R_e}\right] = (2\pi f_r)^2 \qquad (2.32)$$

a form similar to that derived by Minneart (eqn. 2.33) for the natural resonance frequency (f_r) of a bubble of resonance radius, R_r, in a liquid medium of density ρ:

$$f_r = \frac{1}{2\pi R_r}\left\{\frac{3\gamma}{\rho}\left(\frac{P_h + 2\sigma}{R_r}\right)\right\}^{1/2} \qquad (2.33)$$

(NB for large bubbles, in which the surface tension effects are negligible ($P_h \gg 2\sigma/R_r$), equation (2.33) reduces to $f_r = (1/2\pi R_r)[3\gamma P_h/\rho]^{1/2}$, which for a bubble in water ($\rho = 1000$ kg m^{-3}) at 1 atm (1.013×10^5 N m^{-2} and with $\gamma = 1$ may be approximated to $f_r R_r \sim 3$ (with R_r in metres.)

It is important to recognise that not all bubbles are capable of producing significant cavitational effects. The greatest coupling of the ultrasonic energy will occur when the natural resonance frequency (f_r) of the bubble is equal to the applied ultrasonic frequency (f_a). This may be demonstrated by considering the general solution of equation (2.30):

$$r = \frac{P_A}{\rho R_e(\omega_r^2 - \omega_a^2)}\left[\sin\omega_a t - \frac{\omega_a}{\omega_r}\sin\omega_r t\right] \qquad (2.34)$$

which reduces to an indeterminate form at resonance — i.e. when $w_a = w_r$.

Appendix 6 contains a BBC BASIC program which allows you, using equation (2.34), to generate bubble radius-time curves for various values of P_A and R_e. If you use the program you will notice that to keep the computational times in bounds it is necessary to limit the values of P_a, f, etc. For instance, a bubble whose initial radius (R_e) is 8×10^{-5} cm oscillates in a stable though complex manner for several cycles under an applied acoustic frequency of 15 MHz (Fig. 2.15a), yet collapses in less than one cycle if the acoustic frequency is 5 MHz (Fig. 2.15b).

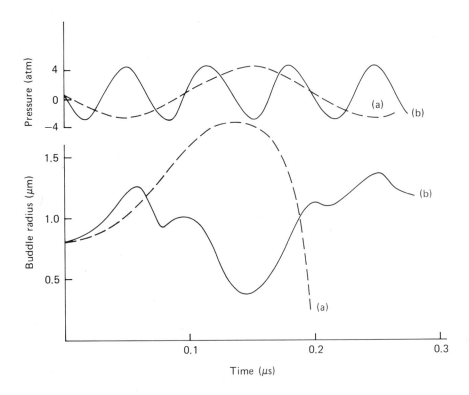

Fig. 2.15 — Radius–time curve for an air bubble in sonicated water at (a) 5 MHz; (b) 15 MHz; $R_e = 8 \times 10^{-5}$ cm; $P_A = 4$ atm; $P_h = 1$ atm.

In other words, the tendency of a bubble to collapse is frequency dependent and is more likely to occur at the lower frequencies where the time available in the compression cycle is longer. Although we will return to this frequency dependence later, you may wish to try using the computer program to verify this fact for yourself. According to equation (2.33), the resonance size of a bubble in water insonated at 20 kHz (a typical sonochemical frequency) is approximately 1.5×10^{-2} cm. If an acoustic pressure amplitude of 1 atm is assumed, then a bubble of radius 10^{-2} cm will collapse (Fig. 2.16), whereas one of radius 2×10^{-2} cm will not (Fig. 2.17).

Further, if an insonation frequency of 50 kHz is employed neither of these two bubbles undergo collapse (Figs. 2.18 and 2.19). Only a bubble close to the resonance size (0.6×10^{-2} cm) will undergo collapse at the higher frequency (Fig. 2.20).

In general for small P_A/P_h ratios, with $R_e \sim R_r$ (the resonant bubble radius), oscillations take place at approximately the excitation frequency (Fig. 2.21). For

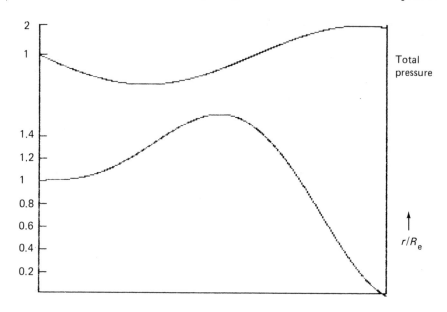

Fig. 2.16 — Radius–time curve for air bubble in sonicated water at 20 kHz. $R_e = 10^{-2}$ cm;
$P_A = 1$ atm; $P_h = 1$ atm.

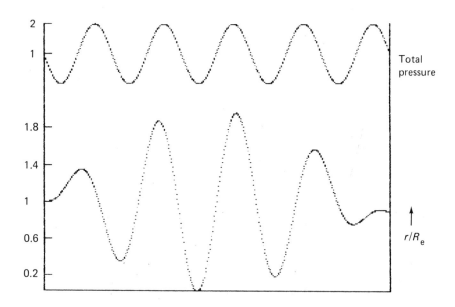

Fig. 2.17 — Radius–time curve for air bubble in sonicated water at 20 kHz. $R_e = 2 \times 10^{-2}$ cm;
$P_A = 1$ atm; $P_h = 1$ atm.

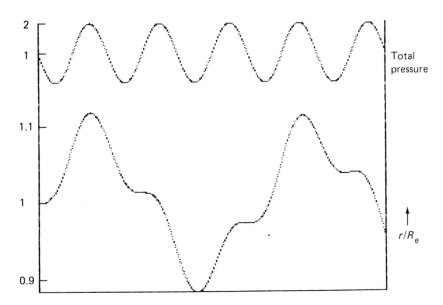

Fig. 2.18 — Radius–time curve for air bubble in sonicated water at 50 kHz. $R_e = 2 \times 10^{-2}$ cm; $P_A = 1$ atm; $P_h = 1$ atm.

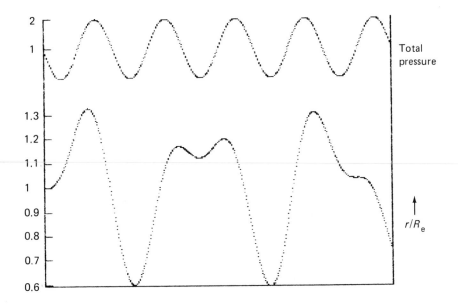

Fig. 2.19 — Radius–time curve for air bubble in sonicated water at 50 kHz. $R_e = 10^{-2}$ cm; $P_A = 1$ atm; $P_h = 1$ atm.

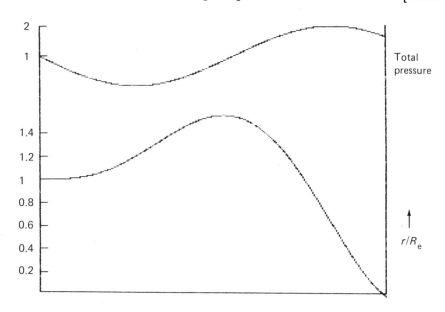

Fig. 2.20 — Radius–time curve for air bubble in sonicated water at 50 kHz. $R_e = 0.5 \times 10^{-2}$ cm;
$P_A = 1$ atm; $P_h = 1$ atm.

$R_e > R_r$ bubble oscillation has a strong component of its own natural resonant frequency (Fig. 2.22). However for very small bubbles, $R_e \ll R_r$, transient conditions are attained as P_A increases beyond P_h (Fig. 2.23, $P_A = 4$ and and 10 atm). It may be that as P_A is further increased the bubble grows so large in the tension phase that it has insufficient time to collapse before the end of the pressure cycle and collapse occurs at the end of the second positive peak (Fig. 2.23, $P_A = 25$, 100, 200 atm). Eventually if $P_A/P_h \gg\gg 1$ the bubble may never undergo transient collapse.

(NB. It will not be possible to faithfully reproduce Fig. 2.23 using the BBC BASIC program in Appendix 6. The program is based upon equation (2.34) where the damping effects of viscosity and temperature upon the bubble wall have been neglected and r has been assumed to be less than R_e.)

What is apparent from Fig. 2.23 is that the fate of the bubble, i.e. whether it remains as a stable bubble or is transformed into a transient, depends upon many factors such as temperature, vapour pressure, hydrostatic pressure, acoustic pressure amplitude (intensity), initial bubble size, viscosity and surface tension. However before discussing the specific effects of some of these parameters let us consider transient and stable cavitation bubbles in more general terms.

2.5.1 Transient cavitation

Transient cavitation bubbles are voids, or vapour-filled bubbles, believed to be produced using sound intensities in excess of 10 W cm^{-2}. They exist for one, or at most a few acoustic cycles, expanding to a radius of at least twice their initial size (Figs. 2.16 and 2.20), before collapsing violently on compression, often disintegrat-

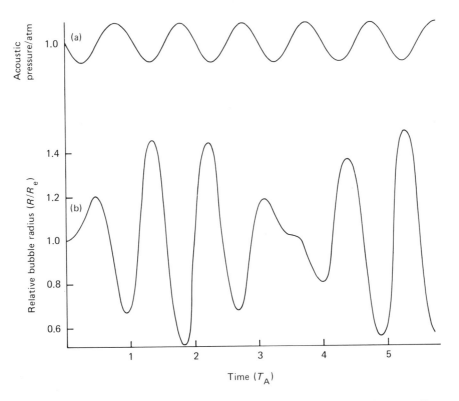

Fig. 2.21 — Radius–time curve for a cavity insonated below resonance frequency R_e = 2.6×10^{-3}cm; P_A = 0.333 atm; P_h = 1 atm. (a) Applied frequency, 83.4 kHz; (b) relative radius of bubble.

ing into smaller bubbles. (These smaller bubbles may act as nuclei for further bubbles, or if of sufficiently small radius (R) they can simply dissolve into the bulk of the solution under the action of the very large forces due to surface tension, $2\sigma/R$). During the lifetime of the transient bubble it is assumed that there is no time for any mass flow, by diffusion of gas, into or out of the bubble, whereas evaporation and condensation of liquid is assumed to take place freely. If there is no gas to cushion the implosion a very violent collapse will result. Theoretical considerations by Noltingk and Neppiras [7] and later by Flynn [8], and separately by Neppiras [9], assuming adiabatic collapse of the bubbles, allows for a calculation of the temperatures (T_{max}) (eqn 2.35) and pressures (P_{max}) (eqn 2.36) within the bubble at the moment of total collapse as being

$$T_{max} = T_0 \left\{ \frac{P_m(K-1)}{P} \right\} \qquad (2.35)$$

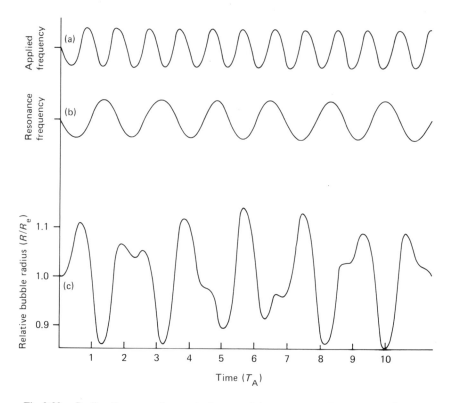

Fig. 2.22 — Radius–time curve for a cavity insonated above resonace. $R_e = 2.6 \times 10^{-3}$ cm; $P_A = 0.333$ atm; $P_h = 1$ atm. (a) Applied frequency, 190 kHz; (b) resonance frequency of bubble; (c) relative radius of bubble.

$$P_{max} = P\left\{\frac{P_m(K-1)}{P}\right\}^{K/(K-1)} \qquad\qquad (2.36)$$

where T_0 is the ambient (experimental) temperature, K is the polytropic index of the gas (or gas vapour) mixture, P is the pressure in the bubble at its maximum size (usually assumed to be equal to the vapour pressure (P_v) of the liquid), and P_m is the pressure in the liquid at the moment of transient collapse. (Those readers who are interested will find a derivation of equations (2.35) and (2.36) in Appendix 5.)

Since the collapse time, τ, for an empty bubble is normally not longer than one-fifth of the period of vibration, P_m can be regarded as constant during the collapse and may be represented as $(P_h + P_a)$. The assumption that the pressure in the bubble may be replaced by the vapour pressure is a direct consequence of the initial assumption that transient bubbles grow without the influx of gas into the cavity. If gas does enter the cavity, the value of $P(= P_g + P_v)$ will depend upon the value of P_g when the bubble is at its maximum size, R_{max}.

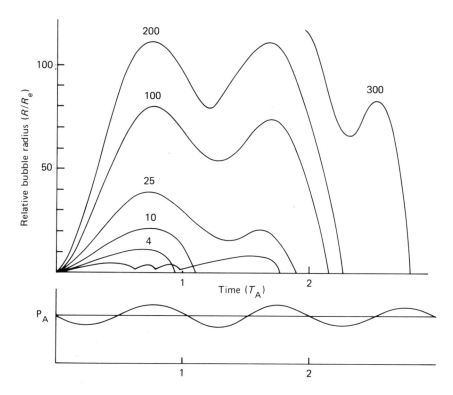

Fig. 2.23 — Radius–time curves for an insonated air bubble in water. $R_e = 10^{-4}$ cm; $f = 500$ kHz. Numbers on the curves refer to the ratio P_A/P_h. Time is measured in units of the period (T_A) of the applied acoustic field.

Using equations (2.35) and (2.36), estimates of the temperatures and pressures involved in the final phase of the implosion can be obtained. For example, for a bubble containing nitrogen $(K = 1.33)$ in water at an ambient temperature of 20°C (T_0) and an ambient pressure of 1 atm (P_m), equations (2.35) and (2.36) provide values of 4200 K and 975 atm respectively. It is the existence of these very high temperatures within the bubble that have formed the basis for the explanation of radical production and sonoluminescence, while the release of the pressure, as a shock wave, is a factor which has been used to account for both increased chemical reactivity (due to increased molecular collision) and polymer degradation.

2.5.2 Stable cavitation
We now turn our attention to stable cavitation, a phenomenon which at one time was not thought to be of much significance in terms of chemical effects. Stable bubbles are thought to contain mainly gas and some vapour and are believed to be produced at fairly low intensities (1–3 W cm^{-2}), oscillating, often non-linearly, about some equilibrium size for many acoustic cycles — Fig. 2.20. The time scale over which they exist is sufficiently long that mass diffusion of gas, as well as thermal diffusion, with consequent evaporation and condensation of the vapour can occur, resulting in

significant long-term effects. If the rates of mass transfer across the gas-liquid interface are not equal, it may result in bubble growth. The mechanism by which small microbubbles in the liquid (which normally dissolve instantly due to surface tension) grow is termed rectified diffusion. In the expansion phase of the acoustic cycle, gas diffuses from the liquid into the bubble, while in the compression phase, gas diffuses out of the bubble into the liquid. Since the interfacial area is greater in the expanded phase, the inward diffusion is greater and leads to an overall growth of the bubble. As the bubble grows the acoustic and environmental conditions of the medium will change and the bubble may be transformed into a transient bubble and undergo collapse (Fig. 2.23). The violence of collapse however, will be less than that for a vapour-filled transient since the gas will cushion the implosion. On the other hand, the bubbles may continue to grow during subsequent cycles until they are sufficiently buoyant to float to the surface and be expelled — this is the process of ultrasonic degassing. As with transient cavitation, estimates have been made of the temperatures and pressures produced in stable bubbles as they oscillate in resonance with the applied acoustic field. Griffing has derived an expression (eqn. 2.37) for the ratio T_0/T_{max}:

$$\frac{T_0}{T_{max}} = \{1 + Q[(P_h/P_m)^{1/3\gamma} - 1] - 1\}^{3(\gamma-1)} \tag{2.37}$$

where Q is the ratio of the resonance amplitude to the static amplitude of vibration of the bubble and P_m ($= P_h + P_a$) is the peak pressure of the bubble. For a bubble containing a monatomic gas ($\gamma = 1.666$) and with $P_m/P_h = 3.7$ (corresponding to an intensity of $2.3\,W\,cm^{-2}$) and assuming a value of $Q = 2.5$, the maximum temperature developed in the bubble (T_{max}) is deduced to be 1665 K. Calculations of the local pressures due to these resonance vibrations has resulted in values which exceed the hydrostatic pressure by a factor of 150 000. There is no doubt that the intense local strains in the vicinity of the resonating bubble are the cause of the many disruptive mechanical effects of sound.

Given the differences in irradiation conditions (i.e. frequency, solvents, vapour pressure, intensity of irradiation, hydrostatic pressure, etc.) it is pertinent to conclude our discussions with a summary of how these various parameters affect the distinct stages of acoustic cavitation: namely nucleation, bubble growth and collapse.

2.6 SUMMARY

2.6.1 Frequency

It is found that as the ultrasonic frequency is increased the production and intensity of cavitation in liquids decreases. Although various proposals have been offered to explain this observation, in qualitative terms at least it may be argued that at very high frequency, where the rarefaction (and compression) cycles are very short, the finite time required for the rarefaction cycle is too small to permit a bubble to grow to a size sufficient to cause disruption of the liquid. Even if a bubble was to be produced during rarefaction, the time required to collapse that bubble may be longer than is available in the compression half-cycle. The resultant cavitational effects, eg shock wave pressure, will therefore be less at the higher frequencies. (Fig. 2.24 shows the variation of maximum fluid-pressure against frequency for constant pressure amplitude (P_A), and bubble radius.)

2.6.2   Solvent
The formation of voids or vapour-filled microbubbles (cavities) in a liquid requires that the negative pressure in the rarefaction region must overcome the natural cohesive forces acting within the liquid. It follows therefore that cavitation should be more difficult to produce in viscous liquids, or liquids with high surface tensions, where the forces are stronger and waves with greater amplitude (P_A) and hence greater intensity will be necessary. Once a liquid produces cavitation bubbles however, the temperature (T_{max}) and pressure (P_{max}) effects resulting from the bubbles collapse will be greater, since the pressure at the start of collapse (P_m) will be larger — see equations (2.35) and (2.36).

Another solvent factor affecting cavitation is that of vapour pressure. Since vapour pressure (P_v) and temperature are directly related this is dealt with below in section 2.6.3.

2.6.3   Temperature
Increasing the reaction temperature allows cavitation to be achieved at lower acoustic intensity. This is a direct consequence of the rise in vapour pressure associated with heating the liquid. The higher the vapour pressure the lower the applied acoustic amplitude (P_A) necessary to ensure that the 'apparent' hydrostatic pressure, $P_h - P_a$, is exceeded — see Section 2.4.4. Unfortunately the effects resulting from cavitational collapse are also reduced. A consideration of equations (2.35) and (2.36) shows that T_{max} and P_{max} fall due to the increase in P_v and decrease in P_m ($= P_h + P_a$). In other words to get maximum sonochemical benefit any experiment should be conducted at as low a temperature as is feasible or with a solvent of low vapour pressure.

2.6.4   Gas type and content
According to equations (2.35), (2.36) and (2.37), employing gases with large γ values will provide for larger sonochemical effects from gas-filled bubbles. For this reason monoatomic gases (He, Ar, Ne) are used in preference to diatomics (N_2, air, O_2). It must be remembered that this dependence on γ is a simplistic view since the extent of the sonochemical effects will also depend upon the thermal conductivity of the gas — i.e. the greater the thermal conductivity of the gas, the more heat (formed in the bubble during collapse) will be dissipated to the surrounding liquid, effectively decreasing T_{max}. Unfortunately a strict correlation between thermal conductivity and sonochemical effect has not been observed (Table 2.2).

Increasing the gas content of a liquid leads to a lowering of both the cavitational threshold (Fig. 2.25) and the intensity of the shock wave released on the collapse of the bubble. The threshold is lowered as a consequence of the increased number of gas nuclei (or weak spots) present in the liquid, whilst the cavitational collapse intensity is decreased as a result of a the greater 'cushioning' effect in the micro-bubble. (The latter point may be deduced semiquantitatively from a consideration of equations (2.35) and (2.36), where P should be replaced by P_{vg} ($= P_v + P_g$).

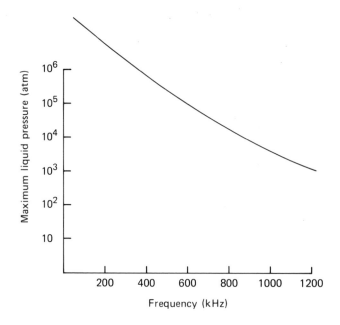

Fig. 2.24 — Variation with frequency of maximum fluid pressure during collapse $R_e = 3.2 \times 10^{-4}$ cm; $P_A = 4$ atm.

Table 2.2 — Rate of formation of free chlorine by irradiation of water containing CCl_4 in relation to the nature of the saturating gas

| Gas | Reaction rate | | Thermal conductivity |
	(mM/min^{-1})	γ	$(10^{-2}\,W\,m^{-1}\,K^{-1})$
Argon	0.074	1.66	1.73
Neon	0.058	1.66	4.72
Helium	0.049	1.66	14.30
Oxygen	0.047	1.39	1.64
Nitrogen	0.045	1.40	2.52
Carbon monoxide	0.028	1.43	2.72

Increasing the gas content of the liquid, P_g, leads to an increase in P_{vg} and a decreases in both P_{max} and T_{max}).

It might be anticipated that employing gases with increased solubility will also reduce both the threshold intensity (by virtue of providing a large number of nuclei in the solvent) and the intensity of cavitation. Indeed there is a definite correlation

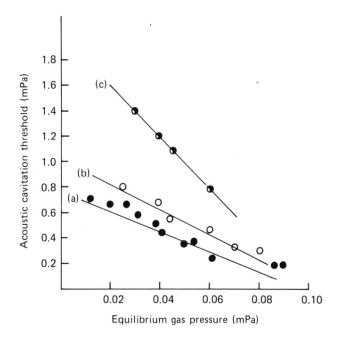

Fig. 2.25 — Variation of acoustic cavitation threshold of water with dissolved gas content. (a) Distilled water, $\sigma = 7.2 \times 10^{-2}\,N\,m^{-1}$. (b) Aqueous guar gum (100 p.p.m.), $\sigma = 6.2 \times 10^{-2}\,N\,m^{-1}$. (c) Aqueous photoflow (80 p.p.m.), $\sigma = 4.0 \times 10^{-2}\,N\,m^{-1}$.

between gas solubility and cavitational intensity. The greater the solubility of the gas the greater the amount which penetrates into the cavitation bubble and the smaller the intensity of the shock wave created on bubble collapse. A further factor affecting the intensity of collapse may be that the more soluble the gas, the more likely it is to redissolve in the medium during the compression phase of the acoustic cycle.

2.6.5 External (applied) pressure
Increasing the external pressure (P_h) leads to an increase in both the cavitation threshold and the intensity of bubble collapse. Qualitatively it can be assumed that there will no longer be a resultant negative pressure phase of the sound wave (since $P_h - P_a > 0$) and so cavitation bubbles cannot be created. Clearly, a sufficiently large increase in the intensity, I, of the applied ultrasonic field could produce cavitation, even at high overpressures, since it will generate larger values of P_a ($I \propto P_A^2$; $P_a = P_A \sin 2\pi ft$) making $P_h - P_a < 0$. In that P_m (the pressure in the bubble at the moment of collapse) is approximately $P_h + P_a$, increasing the value of P_h will lead to a more rapid (eqn 2.27) and violent (eqn 2.36) collapse.

2.6.6 Intensity
In general an increase in intensity (I) will provide for an increase in the sonochemical effects. Cavitation bubbles, initially difficult to create at the higher frequencies (due to the shorter time periods involved in the rarefaction cycles) will now be possible,

and since both the collapse time (eqn. 2.27), the temperature (eqn. 2.35) and the pressure (eqn. 2.36) on collapse are dependent on P_m $(= P_h + P_a)$, bubble collapse will be more violent. However it must be realised that intensity cannot be increased indefinitely, since R_{max} (the maximum bubble size) is also dependent upon the pressure amplitude (eqn. 2.38). With increase in the pressure amplitude (P_A) the bubble may grow so large on rarefaction (R_{max}) that the time available for collapse is insufficient.

$$R_{max} = \frac{4}{3\omega_a}(P_A - P_h)\left(\frac{2}{\rho P_A}\right)^{1/2}\left[1 + \frac{2}{3P_h}(P_A - P_h)\right]^{1/3} \qquad (2.38)$$

As an example to illustrate this point consider the effect of applying an acoustic wave of 20 kHz and pressure amplitude 2 atm to a reaction in water. According to equation (2.38), this amplitude will produce a bubble of maximum radius, R_{max}, 1.27×10^{-2} cm, which if it can be assumed that $P_m = P_a + P_h$; collapses in approximately 6.6 μs (eqn. 2.27). This is less than one-fifth of a cycle (10 μs), the assumption often made for the time interval during which a bubble can undergo transient collapse, and so the bubble could undergo complete collapse. If the 20 kHz wave's amplitude is increased to 3 atm, the maximum size of the bubble can be calculated to be 2.31×10^{-2} cm and the collapse time to be 10.5 μs, i.e. a value greater than one-fifth of a cycle. In other words, this latter bubble does not have sufficient time to undergo transient collapse. Obviously the above is a simplified example and only serves to illustrate the fact that there is a P_A value above which a bubble will have grown so large it will not collapse. This has already been shown graphically in Fig. 2.23.

REFERENCES

[1] A. B. Wood, *A Textbook of Sound*, G. Bell & Son, 1930.
[2] G. G. Stokes, *Trans. Camb. Phil. Soc*, 1849, **8**, 287.
[3] G. Kirchhoff, *Ann. Phys. (Leipzig)*, 1868, **134**, 177.
[4] F. E. Fox and G. D. Rock, *J. Acoust. Soc. Am.*, 1941, **12**, 505.
[5] M. Greenspan and C. E. Tschiegg, *J. Res. Natl. Bur. Stand. Sect. C*, 1967, **71**, 229.
[6] Lord Rayleigh, *Philos. Mag.*, 1917, **34**, 94.
[7] B. E. Noltingk and E. A. Neppiras, *Proc. Phys. Soc. B. (London)*, 1950, **63B**, 674; 1951, **64B**, 1032.
[8] H. G. Flynn, *Physical Acoustics*, Vol 1B, ed. W. P. Mason, Academic Press, New York, 1964, pp. 57–172.
[9] E. A. Neppiras, *Phys. Rep.*, 1980, **61**, 160.

APPENDICES

Appendix 1   Relationship between particle velocity (v) and acoustic pressure (P_a)

Consider an element (DEFG) of a liquid (Fig. A1.1) at a normal pressure of P and density of ρ subject to an applied acoustic wave of pressure P_a.

Under the action of the excess pressure, P_a, the molecules in the element BC will

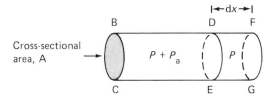

Fig. A1.1

be displaced in the direction of the wave, the particles' velocity increasing from zero at BC to v at DE. The sound wave itself, of velocity c, will pass from DE to FG (i.e. through the distance dx) in dx/c seconds and hence the acceleration (velocity/time) of the particles in this volume element will, on average, be vc/dx, since for such a small element v can be assumed to remain constant.

Obviously the force attained by these accelerating particles (= mass × acceleration) derives from the excess pressure (= pressure × area), such that $(\rho A dx) \times (vc/dx) = (P_a) \times (A)$, i.e. $\rho vc = P_a$ or

$$P_a/v = \rho c \tag{2.11}$$

Appendix 2   Critical pressure (P_K) and bubble radius relationship

If a bubble of radius R_e, present in a liquid is to remain in equilibrium (i.e. neither contracting nor expanding), then the forces (or pressure) acting on the bubble walls attempting to collapse the bubble must equal those forces responsible for attempting to expand the bubble. In the case of expansion this pressure, P_{bub}, will be due to the trapped gas and vapour in the bubble, so that

$$P_{bub} = P_v + P_g \tag{I}$$

whilst for contraction the pressure, P_L, it is the combined effect of the hydrostatic pressure (P_h) within the liquid and the surface tension effect ($2\sigma/R_e$) i.e.

$$P_L = P_h + (2\sigma/R_e) \tag{II}$$

At equilibrium, when neither expansion nor contraction occurs then $P_L = P_{bub}$. Substitution from (I) and (II) gives

$$P_h + (2\sigma/R_e) = P_v + P_g \tag{III}$$

Suppose the bubble now changes size to some new radius, R, as a result of a change in the hydrostatic pressure to a new value P_h'. Then, assuming ideal gas behaviour, the new gas pressure (P_g') inside the bubble will become $P_g(R_e/R)^3$ and the new pressure in the bubble (P_{bub}') will be given by

$$P'_{\text{bub}} \;=\; P_\text{v} + P'_\text{g} \;=\; P_\text{v} + P_\text{g}(R_\text{e}/R)^3 \tag{IV}$$

Since expansion of the bubble to R will decrease the surface tension effects, the new liquid pressure P'_L will be given by:

$$P'_\text{L} \;=\; P'_\text{h} + 2\sigma/R \tag{V}$$

and if the bubble is still to be in equilibrium (i.e. $P'_\text{L} = P'_\text{bub}$ then

$$P'_\text{h} + 2\sigma/R \;=\; P_\text{v} + P_\text{g}(R_\text{e}/R)^3 \tag{VI}$$

or

$$P'_\text{h} \;=\; P_\text{g}(R_\text{e}/R)^3 + P_\text{v} - 2\sigma/R \tag{VII}$$

In other words the dependence of the radius, R, on the hydrostatic pressure of the liquid, P'_h, is not linear but has an inverse cubic dependence, such that for small but constant reductions in P'_h the radius increases quite rapidly. This is especially so when the hydrostatic pressure is quite small. In fact there is a minimum critical hydrostatic pressure, a very small reduction of which causes a dramatic increase in R, i.e. the bubble becomes unstable and grows explosively. The radius at which this occurs may be termed the critical radius, R_K, and occurs when dP'_h/dR is effectively zero (i.e. small change in P'_h, large change in R).

An estimate may be made of this critical bubble radius, R_K, by differentiating the RHS of equation (VII), with respect to R, and equating to zero. That is

$$0 \;=\; -3P_\text{g}\frac{R_\text{e}^3}{R^4} - 0 + \frac{2\sigma}{R^2}$$

Replacement of R by R_K yields

$$R_\text{K}^2 \;=\; \frac{3}{2\sigma} P_\text{g} R_\text{e}^3 \tag{VIII}$$

If the critical pressure at which the bubble attains its critical radius, R_K, is denoted as P_K, then from (VII)

$$P_\text{K} \;=\; P_\text{g}(R_\text{e}/R_\text{K})^3 + P_\text{v} - 2\sigma/R_\text{K} \tag{IX}$$

and replacement of R_K (from VIII) yields

$$P_\text{K} \;=\; P_\text{v} - \frac{2}{3}\left\{\frac{(2\sigma/R_\text{e})^3}{P_\text{g}}\right\}^{1/2} \tag{X}$$

Substitution of P_g $(= P_h - P_v + 2\sigma/R_e)$ from (III) gives

$$P_K = P_v - \frac{2}{3}\left\{\frac{(2\sigma/R_e)^3}{3(P_h - P_v + 2\sigma/R_e)}\right\}^{1/2} \qquad\qquad\text{(XI)}$$

Finally, if the presence of vapour in the bubble is neglected (i.e. $\mathbf{P_v} \sim 0$) then the critical pressure may be given by

$$P_K = -\frac{2}{3}\left\{\frac{(2\sigma/R_e)^3}{3(P_h + 2\sigma/R_e)}\right\}^{1/2} \qquad\qquad\text{(2.18)}$$

The fact that the pressure is negative implies that a negative pressure must be applied to overcome the cohesive forces of a liquid and produce a bubble of radius R_e. Writing $\mathbf{P_K} = P_h - P_B$, allows the estimation of P_B, (known as *Blake threshold pressure*), which is the negative (or rarefaction) pressure which must be applied in excess of the hydrostatic pressure (P_h) to create a bubble of radius R_e. e.g. for large bubbles (i.e. $2\sigma/R_e \ll P_h$)

$$P_B \sim P_h + \frac{8\sigma}{9}\left\{\frac{3\sigma}{2P_h R_e^3}\right\}^{1/2}$$

and for small bubbles $(2\sigma/R_e \gg P_h)$

$$P_B \sim P_h + 0.77\sigma/R_e$$

This latter equation is helpful when attempting to deduce an estimate of the theoretical strength of water.

For example if we assume a void would be created in water when the molecules are separated by more than the van der Waals' distance $(R_e = 4 \times 10^{-10} \text{ m say})$, then taking σ as 76×10^{-3} Nm, P_a can be estimated to be approximately 1500 atmospheres (i.e. P_B $(0.77 \times 76 \times 10^{-4})/(4 \times 10^{-10})$ Nm^{-2}).

Appendix 3   Time of bubble collapse
Consider an empty bubble collapsing under the influence of a constant external pressure P_h from an initial radius R_m to a final radius R (Fig. A3.1).

The work done by the (hydrostatic) pressure, P_h, neglecting surface tension effects, is the product of the pressure and the change in volume and is given by

$$\int_R^{R_m} P_h 4\pi R^2 \mathrm{d}R = P_h\frac{4\pi}{3}[R_m^3 - R^3] \qquad\qquad\text{(XII)}$$

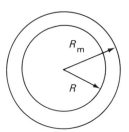

Fig. A3.1

This work will be equal to the kinetic energy ($mv^2/2$) of the liquid as it moves to fill the space vacated by the collapse of the bubble and is given by $1/2 \rho 4 \pi r^2 dr \, (dr/dt)^2$ — Fig. A3.2 where $4 \pi r^2 dr$ = the volume of liquid moving and (dr/dt) is the velocity.

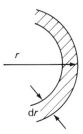

Fig. A3.2

Thus

$$P_h \frac{4\pi}{3}(R_m^3 - R^3) = 2\pi\rho \int_R^\infty r^2 dr (dr/dt)^2 \qquad \text{(XIII)}$$

At first sight this integration appears somewhat difficult to perform since it contains both dr and (dr/dt). However, if it is assumed that the liquid is incompressible then the volume lost by the cavity ($4\pi R^2 dR$) is equal to that filled by the liquid ($4\pi r^2 dr$) — i.e. $4\pi R^2 dR = 4\pi R^2 dR$ or

$$R^2 dR = r^2 dr \qquad \text{(XIV)}$$

Dividing both sides of (XIV) by dt and rearranging gives

$$(dr/dt) = (R^2/r^2)(dR/dt) \tag{XV}$$

Substitution into (XIII) yields

$$\frac{4\pi}{3} P_{\rm h}(R_{\rm m}^3 - R^3) = 2\pi\rho \int_R^\infty r^2 dr \frac{R^4}{r^4}\left(\frac{dR}{dt}\right)^2 = 2\pi\rho R^4\left(\frac{dR}{dt}\right)^2 \int_0^\infty dr/r^2 \tag{XVI}$$

which on integration gives

$$\frac{4\pi}{3} P_{\rm h}(R_{\rm m}^3 - R^3) = 2\pi\rho R^4\left(\frac{dR}{dt}\right)^2\left[-\frac{1}{r}\right]_R^\infty$$

or more precisely

$$\frac{4\pi}{3}\rho P_{\rm h}(R_{\rm m}^3 - R^3) = 2\pi\rho R^3\left(\frac{dR}{dt}\right)^2 \tag{XVII}$$

where the right-hand side is equivalent to the kinetic energy of the liquid. Rearranging gives

$$\left(\frac{dR}{dt}\right) = \left(\frac{2P_{\rm h}}{3\rho}\right)^{1/2}\left(\frac{R_{\rm m}^3}{R^3} - 1\right)^{1/2} \tag{XVIII}$$

such that

$$dt = \frac{dR}{\left[\left(\frac{2P_{\rm h}}{3\rho}\right)\left(\frac{R_{\rm m}^3}{R^3} - 1\right)\right]^{1/2}} \tag{XIX}$$

The time to collapse the bubble can now be obtained by integrating the right-hand side of equation (XIX) from $R_{\rm m}$ to zero to give

$$\tau \approx 0.915 R_{\rm m}(\rho/P_{\rm h})^{1/2} \tag{XX}$$

In the presence of an acoustic field the pressure in the liquid ($P_{\rm m}$) will be greater than the atmospheric pressure by an amount $P_{\rm a}$, i.e. $P_{\rm m} = (P_{\rm h} + P_{\rm a})$ where $P_{\rm a} = P_{\rm A}\sin 2\pi f t$, and equation (XX) may be modified to yield

$$\tau \approx 0.915 R_m (\rho/P_m)^{1/2} \tag{XXI}$$

where P_m is the pressure in the fluid at the start of collapse of a bubble of radius R_m.

The above deduction of τ, although neglecting the effect of both surface tension and the pressure (P_{vg}) of liquid vapour (or gas) most certainly present in the bubble, does allow an estimate of the time involved for collapse.

For example, in the absence of an applied acoustic field ($P_a = 0$), a bubble of radius 10^{-3} cm, (R_m), present in water ($\rho = 10^3$ kg m^{-3}) at 1 atmosphere, will collapse in approximately 1 μs. If this collapse were to occur in the presence of an applied acoustic field, of say 20 kHz (period, $T = 50 \mu$s), then the collapse will occur in a much shorter time than the compression cycle acts (i.e. 25 μs). In other words, P_m ($= P_h + P_a$), the pressure at the start of bubble collapse may be assumed to remain effectively constant over the collapse period since P_a ($= P_A \sin 2\pi f t$) remains effectively constant. Obviously, the greater the intensity (I) of the applied field, the larger is P_A ($= (2\rho c I)^{1/2}$, P_a, and hence P_m, and the faster is the collapse of the bubble. On the other hand, the larger the bubble at the start of collapse (R_m), the longer collapse will take and the less certain is the assumption that P_m remains constant. In fact, if R_m is very large, there may be insufficient time for collapse. Interestingly, a bubble of radius 10^{-3} cm ($\tau \sim 1 \mu$s) will be unable to collapse in an applied acoustic field of 1 MHz (period $= 1 \mu$s) or greater.

Subsequent modifications by Zhoroshev, for the collapse of vapour-filled transients from their maximum radius, R_m, under a liquid pressure of P_m gives

$$\tau = 0.915 R_m (\rho/P_m)^{1/2} (1 + P_v/P_m) \tag{2.27}$$

where $P_v =$ vapour pressure in bubble.

In the absence of an ultrasonic field ($P_m = P_h$), Zhoroshev's expression reduces to the Rayleigh (eqn. (XX)), provided the pressure in the bubble (P_v) is less than that in the liquid (P_h).

Appendix 4   Motion of cavity wall
In Appendix 3, it was shown (eqn (XII)) that the work done by an external pressure, P_h, in collapsing a cavity from a radius of R_m to R, was given by

$$\int_R^{R_m} P_h 4\pi R^2 dR \tag{XII}$$

and that this work was equivalent to the kinetic energy of the liquid moved:

$$KE = 2\pi \rho R^3 \dot{R}^2 \tag{XVII}$$

where $\dot{R} = dR/dt$.

The arguments used, however, referred to an empty cavity and neglected the effects of surface tension (σ).

Consider now the movement of a cavity containing gas and vapour, originally at a radius, R_e, due to an increase in the ambient hydrostatic pressure (P_h) by the application of an acoustic pressure wave (P_a).

At any instant in the compression cycle the new hydrostatic pressure, P'_h ($= P_h + P_a$) will cause the cavity radius to decrease from R_e to R (Fig. A4.1).

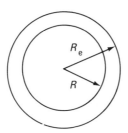

Fig. A4.1

This collapse will be augmented by an increase in the surface tension effect ($2\sigma/R$) as the cavity becomes smaller, i.e the total 'collapse' pressure is ($P'_h + 2\sigma/R$), but will be opposed by the increase in the pressure within the bubble due to the compression of gas, i.e. expanding pressure, P'_{bub}. By analogy with the empty cavity, the work done by the new hydrostatic pressure (P'_h), minus that of the layer adjacent to the bubble, is equal to the kinetic energy of the liquid.

$$-\int_{R_e}^{R} [(P'_h + \frac{2\sigma}{R} - P'_{bub})4\pi R^2 dR = 2\pi\rho R^3 \dot{R}^2 \qquad \text{(XXII)}$$

Differentiation of (XIII) yields

$$\left(P'_{bub} - P'_h - \frac{2\sigma}{R}\right)4\pi R^2 dR = 2\pi\rho[3R^2\dot{R}^2 dR + R^3 2\ddot{R} dR] \qquad \text{(XXIII)}$$

which on dividing by $2\pi R^2 dR$ and rearranging gives

$$R\ddot{R} + \frac{3}{2}\dot{R}^2 = \frac{1}{\rho}[P'_{bub} - P'_h - 2\sigma/R] \qquad \text{(XXIV)}$$

In Appendix 2, is was shown that

$$P'_{bub} = P_v + P_g \left(\frac{R_e}{R}\right)^3 \qquad \text{(IV)}$$

and

$$P_g = \left(P_h + \frac{2\sigma}{R_e}\right) - P_v)$$

(III)

Substitution into eqn (XXIV), and replacement of P_h' by $(P_h + P_a)$, gives, on neglecting vapour pressure ($P_v = 0$), equation (XXV)

$$RR'' + \frac{3}{2}\dot{R}^2 = \frac{1}{\rho}\left[\left(P_h + \frac{2\sigma}{R_e}\right)\left(\frac{R_e}{R}\right)^3 - \frac{2\sigma}{R} - P_h - P_a\right]$$

(XXV)

If the acoustic pressure, P_a, is replaced with $-P_A \sin \omega_A t$ (the form appropriate for a sinusoidal wave of pressure amplitude P_A and circular frequency $w_A(= 2\pi f)$, equation (XXV) reduces to that first derived by Noltingk and Neppiras.

$$RR'' + \frac{3}{2}\dot{R}^2 = \frac{1}{\rho}\left[\left(P_h + \frac{2\sigma}{R_e}\right)\left(\frac{R_e}{R}\right)^3 - \frac{2\sigma}{R} - P_h + P_A \sin \omega_A t]$$

More strictly the equation is

$$RR'' + \frac{3}{2}\dot{R}^2 = \frac{1}{\rho}\left(P_h + \frac{2\sigma}{R_e}\right)\left(\frac{R_e}{R}\right)^{3K} - \frac{2\sigma}{R} - P_h + P_A \sin \omega_A t\right]$$

(2.29)

where K is the polytropic index of the gas and varies between γ, the specific heat ratio, and unity, the limits for adiabatic and isothermal conditions.

Appendix 5   Maximum bubble temperature (T_{max}) and pressure (P_{max})
In Appendix 4 we derived the equation for the motion of a cavity wall, as it collapsed due to an external pressure P_h', to be

$$RR'' + \frac{3}{2}\dot{R}^2 = \frac{1}{\rho}\left[P_{bub}' - P_h' - \frac{2\sigma}{R}\right]$$

(XXIV)

where P_{bub}', the pressure inside the bubble at a radius of R, was given by

$$P_{bub}' = P_v + P_g\left(\frac{R_e}{R}\right)^3 \quad (eqn\ IV,\ Appendix\ 2)$$

(P_g = gas pressure in a bubble of initial radius R_e).
   It has been argued (Appendix 3, eqn (XXI)) that the collapse time for a bubble, initially of radius R_e, is considerably shorter than the time period of the compression

cycle. Thus the external pressure P_h' ($= P_a + P_h$), in the presence of an acoustic field, may be assumed to remain effectively constant (P_m) during the collapse period. Neglecting surface tension and vapour pressure effects, assuming adiabatic compression (i.e. very short compression time), and replacing R_e, by R_m, the size of the bubble at the start of collapse, the motion of the bubble wall becomes

$$R\ddot{R} + \frac{3}{2}\dot{R}^2 = \frac{1}{\rho}\left[P_g\left(\frac{R_m}{R}\right)^{3\gamma} - P_m \right] \tag{XXIVa}$$

Unlike Rayleigh's original example of a collapsing empty cavity, this bubble will reduce to a minimum size, R_{min}, on compression, after which it will expand to R_m and subsequently it will oscillate between the two extremes R_m and R_{min}. Obviously at the two extremes of radii, motion of the bubble wall is zero — i.e. $\dot{R} = 0$. To determine these radii it is necessary to integrate (XXIVa).

With $Z = (R_m/R)^3$, the integration yields

$$\frac{\rho\dot{R}^2}{2} = P_m(Z - 1) - \left[\frac{P_g(Z - Z^\gamma)}{(1 - \gamma)} \right] \tag{XXVI}$$

which on setting $\dot{R} = 0$ and rearranging gives

$$P_m(Z - 1)(\gamma - 1) = P_g(Z^\gamma - Z) \tag{XXVII}$$

For very small values of R, (i.e. R_{min}), Z will be extremely large and $(Z - 1)$ approximates to Z, such that equation (XXVII) may be written as

$$P_m(\gamma - 1) \sim P_g Z^{\gamma - 1} \tag{XXVIII}$$

which on rearrangement yields

$$Z = \left(\frac{P_m(\gamma - 1)}{P_g} \right)^{1/(\gamma - 1)} \tag{XXIX}$$

Now at minimum cavity volume (V_{min}) the gas will have its maximum pressure value, P_{max}, such that

$$P_{max}V_{min}^\gamma = P_g V_{max}^\gamma \tag{XXX}$$

And, since the volume of the cavity (V) is related to its radius (R) by

$$V = \frac{4}{3}\pi R^3$$

then

$$\frac{V_{max}}{V_{min}} = \left\{\frac{R_m}{R_{min}}\right\}^3 = Z \qquad\qquad (XXXI)$$

and

$$\left\{\frac{V_{max}}{V_{min}}\right\}^\gamma = \frac{P_{max}}{P_g} = Z^\gamma \qquad\qquad (XXXII)$$

Replacement of Z from (XXIX) and rearrangement gives

$$P_{max} = P_g \left\{\frac{P_m(\gamma-1)}{P_g}\right\}^{\gamma/(\gamma-1)} \qquad\qquad (2.36)$$

At the instant of bubble collapse, this pressure will be released into the liquid. It is these very high liquid pressures which give rise to certain of the well known effects of ultrasonic irradiation, such as erosion, dispersion and molecular degradation.

To find the maximum gas temperature attained at V_{min}, we apply

$$T_{max}V_{min}^{\gamma-1} = T_{min}V_{max}^{\gamma-1}$$

i.e.

$$T_{max} = T_{min}\left\{\frac{P_m(\gamma-1)}{P_g}\right\} \qquad\qquad (2.35)$$

In most sonochemical applications T_{min} is taken to be the ambient (experimental) temperatures, and P_g as the vapour pressure of the liquid at that temperature.

Thus, even if a bubble does not undergo transient collapse, extremely high temperatures and pressures are developed within the bubbles as they oscillate from R_m to R_{min}. It is these large temperatures and pressures within the bubble which are thought to contribute to the significant increases in chemical reactivity observed in the presence of ultrasound. This does suppose, however, that the reactant species is sufficiently volatile to enter the bubble. If it is not, increased chemical reactivity can only be assumed to have occurred due to increased molecular interaction within the liquid due to the large build-up of pressure at the bubble wall which, if the bubble is not to completely collapse, will be equal to the liquid pressure at the liquid–bubble wall interface.

Appendix 6   Bubble dynamics

This program has been designed to run on both an Electron and any BBC model of computer. You will also need

A disk drive
A monitor (preferably colour)
A printer

Operation of the program

(1) *Lines* 20 *to* 310
In order that the program can calculate both the resonance frequency (RF) and angular resonance frequency (WR) of your bubble (radius, br) you will need to input the solvent density (d), the surface tension of the solvent (s), the solvent viscosity (v) and the solvent vapour pressure (pv). The computer will correct the data to the appropriate SI unit. For example, if you wish to enter a bubble radius of 10^{-2} cm you should respond to the question 'Type in the initial (equil) bubble radius in cm' — line 280 — by entering either 0.01 or 1E-2 and pressing return. The computer will convert this value to metres — (i.e. BR = 0.01∗br)

(2) *Lines* 320 *to* 410
This section of the program allows an estimation of the acoustic amplitude (P_A) of the wave from a knowledge of the intensity (Int) of the wave and the velocity (vel) of sound in the medium.

When entering the value for velocity you may type out the number in full (i.e. 1500) or use the mathematical notation 1.5E3.

(3) *Lines* 410 *to* 640
This section of the program determines the maximum and minimum acoustic pressure and bubble radius so that on displaying the data maximum use of the screen is ensured.

(4) *Lines* 650 *to* 1250
This section of the program calculates and displays the variation in the applied acoustic pressure, the resonance pressure (or frequency) and the bubble dimensions with time. If you do not have a coloured monitor change the GCOL statements in lines 1200 and 1220 from GCOL0,1 (red) to GCOL0,3 (white).

Once you have typed the program you should save it as 'BUBBLE'. You may of course choose any other title; however if you do so you will have to amend line 1270.

```
   10 GOSUB1290:PRINTTAB(12,5);"BUBBLE DYNAMICS"''TAB(8,8)"Copyright J P Lorimer
"''TAB(14,10)" Jan 1988":FOR Q=1TO15000:NEXT
   20 GOSUB1290:N=0:GCOL0,3:CLS:PRINT''':INPUT"  Is the solvent water.Y/N";ANS$
   30 IF ANS$="Y" THEN 40 ELSE 50
   40 D=1000:S=.076:V=.001:PV=(.023*1.013E5):GOTO250
   50 GOSUB1300:INPUT" Input solvent density";d:GOSUB1300:PRINT " Are the densit
y units"''" (1) g/cc "''" (2) Kg/m3":GOSUB1310
   60 IF ANS=1 THEN 70 ELSE 80
   70 D=1000*d:GOTO 90
   80 D=d
   90 GOSUB1300:INPUT" Input solvent surface tension";s:GOSUB1300:PRINT " Are th
e units "''" (1) dyne/cm "''" (2) N/m":GOSUB1310
  100 IF ANS=1 THEN 110ELSE 120
  110 S=.001*s:GOTO 130
  120 S=s
  130 GOSUB1300:INPUT" Input solvent viscosity";v:GOSUB1300:PRINT " Are the unit
s "''" (1) centipoise "''" (2) poise "''" (3) Kg/m/s":GOSUB1310
  140 IF ANS=1 THEN 150ELSE 160
  150 V=.001*v:GOTO 190
  160 IF ANS=2 THEN 170 ELSE 180
  170 V=.1*v:GOTO 190
  180 V=v
  190 GOSUB1300:INPUT" Input solvent vapour pressure";pv:GOSUB1300:PRINT " Are t
he units"''" (1) Torr"''" (2) bar"''" (3) N/sq m":GOSUB1310
  200 IF ANS=1 THEN 210 ELSE 220
  210 PV=(pv/760)*1.013E5:GOTO 250
  220 IF ANS=2 THEN 230 ELSE 240
  230 PV=(pv*1.013E5):GOTO 250
  240 PV=pv
  250 REM ATMOS PRESSURE TAKEN TO BE 1 BAR
  260 P0=1.013E5
  270 GOSUB1300:INPUT" Type in the exptl freq in KHz";f:W=1E3*f*2*3.142
  280 GOSUB1300:INPUT"   Type in the initial (equil) bubble"''"   radius in cm
";br:BR=.01*br:IF br<1.01E-2 THEN G=1 ELSE G=1.333
  290 REM - CALCULATION OF RESONANCE FREQ
  300 a=(P0+(2*S/BR)-PV):b=(2*S/BR):c=((4*V*V)/(D*BR*BR)):e=(3*G*a)-b-c:WR=((1/(
D*BR*BR))*e)^.5:RF=WR/(2*3.142)
  310 REM to calculate r value for various PA and t
  320 GOSUB1300:INPUT" Do you know intensity of sound wave";ANS$
  330 IF ANS$="Y" THEN 340 ELSE 360
  340 INPUT "Input Intensity in W/sq cm";Int
  350 GOS1610:INPUT" Input velocity of sound (m/s) in your medium";vel:PA=(2*Int
*D*vel*1E4)^.5:GOTO 420
  360 GOSUB1300:INPUT " Input pressure amplitude (PA) value";pa:GOSUB1300:PRINT
" Are the units"''" (1) Torr"''" (2) bar"''" (3) N/m2":GOSUB1310
  370 IF ANS=1 THEN 380 ELSE 390
  380 PA=(pa/760)*1.013E5:GOTO 420
  390 IF ANS=2 THEN 400 ELSE 410
  400 PA=pa*1.013E5:GOTO 420
  410 PA=pa
  420 ff=((WR^2)-(W^2))
  430 CONST=(PA/(D*BR*ff))
  440 GOSUB1300:INPUT" How many cycles (MAX 10)";N
  450 IF N>10 THEN N=10 ELSE N=N
  460 FC=500/N:INC=1/(FC*f*1E3):T=0:T=T-INC
  470 GOSUB1300:INPUT" Do you want res freq on graph";ANS$
  480 RMAX=1:RMIN=1:PMAX=1:PMIN=1:GOSUB1300
  490 VDU5:MOVE150,500:PRINT"LOOKING FOR MAX AND MIN PRESSURES":VDU 4
  500 FOR I= 1 TO 500
  510 GOSUB1320
  520 IF RR>0.01*BR THEN 530 ELSE 630
  530 C=C+1
  540 IF RR>RMAX THEN 550 ELSE 560
  550 RMAX=RR
  560 IF RR<RMIN THEN 570 ELSE 580
  570 RMIN=RR
  580 IF P>= PMAX THEN 590 ELSE 600
  590 PMAX=P
  600 IF P<PMIN THEN 610 ELSE 640
  610 PMIN=P
  620 GOTO 640
  630 I=500
  640 NEXT
  650 NEWN=(N*C)/500:N=NEWN:GOSUB 1290:IF PMIN>=0 THEN 660 ELSE 670
  660 AA=800:GOTO680
```

```
 670 AA=900
 680 XS=(1000-AA)/PMAX:NN=(INT((ABS(PMAX)+ABS(PMIN))/(2*PO/1.013E5)))+1
 690 IF ANS$="Y" THEN 700 ELSE 710
 700 BB=600:GOTO 720
 710 BB=800
 720 YS=1000/N:FC=500/N:RS=(BB-200)/(RMAX-RMIN):INC=1/(FC*f*1E3):T=-INC
 730 TO=1/(f*1E3)
 740 FOR I= 1 TO (FC*N)
 750 GOSUB1320
 760 t=T/TO:TT=t*YS:PP=(P*XS)+AA:RRR=((RR-RMIN)*RS):RRP=(XS*RP)+BB:@%=&50405
 770 VDU 5:GCOLO,3:MOVE 100,100:DRAW100,1000
 780 IF ANS$="Y" THEN 790 ELSE 800
 790 GCOL 0,2:PLOT69,(TT+100),RRP
 800 GCOL 0,3:PLOT69,(TT+100),PP
 810 GCOL 0,1:PLOT69,(TT+100),RRR+100:VDU4
 820 NEXT   I
 830 GCOLO,3:VDU5:MOVE TT+100,100:DRAW TT+100,1000:MOVE 100,100:DRAW TT+100,100
:VDU 4
 840 IF NN<3 THEN 850 ELSE 860
 850 PAP=1:GOTO 910
 860 IF NN>3 AND NN<5 THEN 870 ELSE 880
 870 PAP=.5:GOTO 910
 880 IF NN>5 AND NN<10 THEN 890 ELSE 900
 890 PAP=.25:GOTO 910
 900 IFNN>10 AND NN<20 THEN PAP=.1 ELSE PAP=.05
 910 IP=-NN+1
 920 FOR I= 1 TO (2*PAP*NN)
 930 IP=IP+(1/PAP):IPP=(IP*XS)+AA:GCOL 0,3:VDU5:MOVE 100,IPP:DRAW 120,IPP:VDU 4
 940 IF ANS$="Y" THEN 950 ELSE 960
 950 VDU5:MOVE 100,(IPP+BB-AA):DRAW 60,(IPP+BB-AA):VDU4
 960 VDU5:MOVE 20,IPP+10:PRINT" ";IP:VDU4
 970 IF ANS$="Y" THEN980 ELSE 990
 980 GCOLO,2:VDU5:MOVE -20,IPP+10+BB-AA:PRINT" ";IP:VDU4
 990 NEXT
1000 IF RMAX-RMIN>0 AND RMAX-RMIN<.8 THEN 1010 ELSE 1020
1010 AB=.1:GOTO 1130
1020 IF RMAX-RMIN>0.8 AND RMAX-RMIN<1.6 THEN 1030 ELSE 1040
1030 AB=.2 :GOTO 1130
1040 IF RMAX-RMIN>1.6 AND RMAX-RMIN<3.3 THEN 1050 ELSE 1060
1050 AB=.4:GOTO 1130
1060 IF RMAX-RMIN>3.3 AND RMAX-RMIN<6.5 THEN 1070 ELSE 1080
1070 AB=.8:GOTO 1130
1080 IF RMAX-RMIN>6.5 AND RMAX-RMIN<9 THEN 1090 ELSE 1100
1090 AB=1:GOTO 1130
1100 IF RMAX-RMIN>9 AND RMAX-RMIN<20 THEN 1110 ELSE 1120
1110 AB=2:GOTO 1130
1120 IF RMAX-RMIN>20 AND RMAX-RMIN<40 THEN AB=4 ELSE AB=8
1130 D=INT((RMAX-RMIN)/AB)+1:DD=-(D*AB):NR=DD
1140 FOR I= 1 TO (2*D)
1150 DD=DD+AB:NR=NR+AB:Sr=DD:BR:R=BR+Sr:RR=R/BR:IIR=((RR-RMIN)*RS):IIRMAX=((RMA
X-RMIN)*RS):IIRMIN=((RMIN-RMIN)*RS)
1160 IF IIR>IIRMAX THEN 1190 ELSE 1170
1170 IF IIR<IIRMIN THEN 1190 ELSE 1180
1180 GCOLO,1:VDU5:MOVE100,IIR+100:DRAW120,IIR+100:MOVE0,IIR+120:PRINT"";NR+1:VD
U4
1190 NEXT
1200 GCOLO,1:VDU5:MOVE 250,70:PRINT"Radius-time curve":VDU4
1210 GCOLO,3:VDU5:MOVE 30,30:PRINT"PA=";PA/1.013E5;" atmos";" f=";f;" kHz";" RO
=";br;" cm":VDU4
1220 GCOLO,1:VDU5:MOVE TT+150,300:PRINT"r":MOVE TT+150,270:PRINT"-":MOVE TT+150
,240:PRINT"Re":MOVE TT+150,400:PRINT"^":MOVE TT+150,370:PRINT"!":GCOLO,3:VDU4
1230 IF ANS$="Y" THEN 1240 ELSE 1250
1240 GCOLO,2:VDU5:MOVE TT+120,BB+100::PRINT"reson":MOVE TT+120,BB+50:PRINT"pres
s":VDU4
1250 GCOLO,3:VDU5:MOVE TT+120,AA+100::PRINT"appl":MOVE TT+120,AA+50:PRINT"press
":VDU4
1260 GCOLO,3:VDU5:MOVE300,800:INPUT" RUN AGAIN.Y/N ";ANS$
1270 IF ANS$="Y" THEN CHAIN"BUBBLE"
1280 END
1290 MODE 1:GCOL 0,129:VDU 19,0,4;0:CLG:GCOL 0,3:RETURN
1300 CLS:PRINT'''':RETURN
1310 PRINT:INPUT" Insert the appropriate number";ANS:RETURN
1320 T=T+INC:P=(PO-(PA*SIN(W*T)))/1.013E5:RP=(PO-(PA*SIN(WR*T)))/1.013E5:r=CONS
T*(SIN(W*T)-((W/WR)*SIN(WR*T))):R=BR+r:RR=R/BR:RETURN
```

3

Synthesis

For the majority of synthetic chemists interest in sonochemistry will be in power ultrasound — the type of sound which provides sufficient energy to affect chemical reactivity (see Chapter 1).

When one wishes to change the reactivity of a system the method of choice would normally be an alteration in one of the external physical variables such as heat, light or pressure. The choice will be made based upon one's experience of the type of effect induced by such alterations. Ultrasound provides energy in a form which is different from the above. The affects of ultrasound however are not as quantifiable as those produced by the more conventional physical variables although the underlying reason for its mode of action is recognised as being due to cavitation.

We have already seen how cavitation can be induced in any liquid provided the appropriate ultrasonic energy is applied and that cavitation is the source of the dramatic effect of power ultrasound on chemical reactivity. A number of external parameters influence the efficacy of ultrasonic irradiation in promoting cavitation and thus chemical reactivity and these have been discussed in detail in Chapter 2. Here we will re-examine briefly the experimental parameters which are more commonly varied. As the frequency is increased cavitation is reduced. At high frequency (>3 MHz) there is not enough time during the rarefaction cycle for bubble growth to achieve sufficient size to disrupt the liquid. For cavitation at high frequency more power is needed since more energy is lost to molecular motion of the liquid. It is for this reason that most commercially available ultrasonic equipment used in sonochemistry operates at the lower end of the ultrasonic range (20–50 kHz).

The first requirement for the level of ultrasonic power required to cause chemical effects in a reaction is that sufficient acoustic energy must be supplied to overcome the cavitation threshold of the medium. Once this has been exceeded then the region of cavitation around the radiating source, the 'cavitational zone', will increase with intensity and one might expect that the sonochemical rate would also increase in parallel with this. Thus the observed increase in the rate of hydrolysis of methyl ethanoate in the presence of ultrasound has been found to be directly proportional to intensity. A limiting value is reached however and beyond this limit the sonochemical rate is found to decrease with power [1,2]. The suggested explanation for this decrease is that with increasing intensity of the sound wave bubble growth during the rarefaction cycle becomes so great that there is insufficient time for bubble collapse. An additional factor to be considered when a liquid is under intense irradiation is a sound 'dampening' effect which reduces power dissipation to the system. This effect is caused when a large number of bubbles are generated around the irradiation source which acts as a barrier to direct transfer of sound energy through the medium.

When considering the reaction conditions for a sonochemical process the choice of solvent and the bulk working temperature are significant factors and are often interrelated. Any increase in solvent vapour pressure would decrease the maximum bubble collapse temperature and pressure. Hence for a reaction where cavitational collapse is the primary cause of sonochemical activation a low bulk temperature would be preferred, particularly if a low boiling solvent is used. Conversely, for a reaction requiring elevated temperatures a high boiling solvent would be appropriate. For heterogeneous reactions where the effect of sonication is action on the surface of either a catalyst or an inorganic solid a balance must be struck between ensuring enough cavitation to achieve reagent activation without overly disturbing the thermodynamics of the reaction. It must also be remembered that in some cases where extended reaction periods are required the solvent itself may not be totally inert, discoloration and charring may occur. In most synthetic reactions however solvent reaction can be ignored because reaction times are short.

Application of an external pressure to a reaction system, which increases the hydrostatic pressure of the liquid, leads to an increase in the energy required to initiate cavitation. In practical terms if this threshold energy can be exceeded with the available irradiation source then an increase in hydrostatic pressure would lead to an increase in sonochemical effect. This is because the maximum temperatures and pressures experienced during bubble collapse will be higher under these conditions. Conversely, bubbling a gas through a liquid during ultrasonic irradiation should decrease both the cavitational threshold and intensity. This is a direct consequence of the increase in the number of gas nuclei available to initiate cavitation and the 'cushioning' effect on the collapse of the bubbles caused by the gas present in them.

The precise way in which ultrasonic energies are transmitted to the chemical reaction is not however straightforward since there are a number of possible sites for reaction depending on the type of system under sonication (Table 3.1).

Homogeneous liquid media

It is tempting to attribute ultrasonically enhanced chemical reactivity (particularly in heterogeneous systems) to the mechanical results of cavitational bubble collapse. That this cannot be the whole reason for the effect of ultrasound on reactivity is clear when we turn our attention to homogeneous reactions. In the sonication of water for example how can we explain the production of radical species and sonoluminescence (the emission of light)?

The answer to these questions lies in the actual process of cavitational collapse. The microbubble is not enclosing a vacuum — it contains vapour from the solvent and any volatile reagents so that, on collapse, these vapours are subjected to the enormous increases in both temperature and pressure referred to above. Under such extremes the solvent and/or reagent vapour suffers fragmentation to generate

Table 3.1 — Possible sites of reaction induced by cavitation

(1) *In homogeneous liquid media*

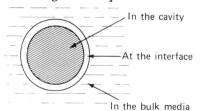

In the cavity — Vapour attains high pressures and temperatures during collapse

At the interface — Shock wave on collapse plus concentration of reactive species

In the bulk media — Shock wave on collapse

(2) *At solid/liquid interface*
Powders

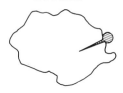

Trapped gas on surfaces and in defects cause nucleation and collapse on surface

Surface cleaning and fragmentation
Solid surfaces

Collapse near solid surface in the liquid phase causes microjet to hit surface
Surface erosion and cleaning

(3) *At a liquid/liquid interface*

Disruption of phase boundary
Highly efficient emulsification

reactive species of the radical or carbene type some of which would be high enough in energy to fluoresce. These high-energy species would be concentrated at the interface and thus give rise to intermolecular reactions, e.g. radical coupling.

If there were involatile solutes present in the homogeneous reaction we might anticipate that these would collect at this same interface. As a consequence these would also be subject to extreme conditions on bubble collapse. They could well react with the high-energy species generated in the vapour phase.

In addition to these effects the shock wave produced by bubble collapse (or even molecular motion induced by the propagating ultrasonic wave itself), could act to disrupt solvent structure. This could then influence reactivity by altering the solvation of the reactive species in the immediate neighbourhood of the disturbance.

Solid/liquid interface

Reactions involving metals

There are two types of reaction involving metals (i) in which the metal is a reagent and is consumed in the process and (ii) in which the metal functions as a catalyst. It is tempting to explain that the ultrasonically induced enhancements in chemical reactivity which are observed in such heterogeneous reactions are due simply to the well-known cleaning action of ultrasound. It is certainly true that dirty surfaces can inhibit chemical reactions involving metals and that sonication will clean the surface of a metal. It is because of surface contamination that many of the metals used in chemical reactions are cleaned before use — e.g. copper is washed with EDTA to remove surface salts and iodine is commonly used in the preparation of a Grignard reagent to remove oxide film and promote magnesium reactivity. In some respects sonication serves a similar purpose to both of these chemical techniques — it exposes clean or reactive surface to the reagents involved. Examination of irradiated surfaces by electron microscopy reveals 'pitting' of the surface of the metal which acts both to expose new surface to the reagents and increase the effective surface area available for reaction [3]. The pitting is thought to be caused by two possible processes: (i) the implosion of cavitation bubbles formed from seed nuclei on the surface or (ii) microstreaming of a jet of solvent onto the surface when a cavitation bubble collapses in the solvent close to it.

In many cases however it has been shown that the cleaning effect alone is insufficient to explain the extent of the sonochemically enhanced reactivity. In such cases it is thought that sonication serves to sweep reactive intermediates, or products, clear of the metal surface and thus present renewed clean surface for reaction. This sweeping effect would not be so effective under normal mechanical agitation [4].

Reactions involving powders or other particulate reagents

In heterogeneous reactions involving solids dispersed in liquids the overall reactivity, just as with the metal surface reactions described above, will depend upon the available reactive surface area. The difference when using powders (metallic or non-metallic) is that ultrasonic 'pitting' can lead to fragmentation and consequent particle size reduction. One interesting aspect of such reductions is that for a particular set of experimental conditions there appears to be an optimum size for the reduction beyond which ultrasound has no further effect [4].

One important benefit of particle size reduction and simultaneous surface activation is the possibility of substitution of sonication for the use of a phase transfer catalyst (PTC) as a means of assisting heterogeneous solid/liquid reactions.

Liquid/liquid interface

Ultrasound is known to generate extremely fine emulsions from mixtures of immiscible liquids. One of the main consequences of these emulsions is the dramatic increase in the interfacial contact area between the liquids, i.e. an increase in the region over which any reaction between species dissolved in the liquids can take place. As with the powder reactions this can result in the use of ultrasound in place of a PTC. In some cases however it has been found that a combination of sonication and PTC has a better overall effect than either of the two techniques alone.

3.1 HOMOGENEOUS SONOCHEMISTRY

3.1.1 Using an Ultrasonic Bath

Ultrasonic irradiation has been used to substantially improve the rates of reaction and product yields of a wide range of reactions but there have been very few detailed investigations of the kinetics of ultrasonic acceleration. One of the exceptions to this has been studies of a range of homogeneous hydrolysis reactions in aqueous solution with the reaction vessel immersed in an ultrasonic bath. In 1968 Folger and Barnes [2] reported that ultrasonic irradiation increased the rates of hydrolysis of methyl ethanoate and attributed the increase in reaction rate to the high temperatures reached within the cavitation bubbles (eqn 3.1). Subsequent studies of the same system by other groups suggested that the increase in reaction rate might also be due to an increase in frequency of collisions between molecules caused by cavitation pressure gradient and temperature rise [1,5]. Studies on the effect of temperature on the increased rate induced by irradiating the system at 540 kHz revealed a 28% enhancement at 20°C and yet only a 20% enhancement at 30°C. This inverse relationship between ultrasonically induced rate acceleration and reaction temperature is now recognised as normal behaviour for systems under irradiation. It is the direct result of the lowering in solvent vapour pressure with a fall in temperature of the system *vide supra*.

$$CH_3COOCH_3 \;+\; H_2O \;\longrightarrow\; CH_3COOH \;+\; HOCH_3 \qquad (3.1)$$

In 1981 Kristol reported ultrasonically induced rate enhancements for the hydrolysis of the 4-nitrophenyl esters of a number of aliphatic carboxylic acids (1, R=Me, Et, *i*-Pr, *t*-Bu) (eqn 3.2) [6]. The rate enhancements at 35°C were all in the range of 14–15% and were independent of the alkyl substituent (R) on the carboxylic acid. It was concluded that the rate enhancements could not be due simply to any increase in the macro reaction temperature (due to the bulk heating effect of the ultrasonic irradiation) since there are marked differences in the energy of activation (E_a) for the hydrolyses of each of these substrates. Given such differences in activation energy, any bulk heating effect (which would be similar in magnitude for each system studied) must result in widely different rate enhancements and not the uniform value of *ca.* 15% observed throughout.

$$R\,COO \!-\!\langle\!\bigcirc\!\rangle\!-\! NO_2 \;\longrightarrow\; R\,COOH \;+\; HO\!-\!\langle\!\bigcirc\!\rangle\!-\! NO_2 \qquad (3.2)$$

(1)

Ultrasound has been found to influence the homogeneous hydrolysis of 2-chloro-2-methylpropane in aqueous alcoholic media (eqn 3.3). This system has been the subject of numerous kinetic studies since it is one of the classic examples of a unimolecular nucleophilic displacement reaction (termed S_N1).

$$t\text{-Bu}-\text{Cl} \quad + \quad \text{R-OH} \quad \rightleftharpoons \quad t\text{-Bu}^+ /\!\!/ \text{Cl}^- \quad \longrightarrow \quad t\text{-BuOR} \quad + \quad \text{HCl} \qquad (3.3)$$

The homogeneous solvolysis of this substrate in aqueous ethanolic solvents can be monitored by the change in conductance as HCl is produced. Initial studies of the reaction in aqueous ethanol as solvent at 25°C using a cleaning bath (45 kHz) revealed modest rate enhancements (up to about two-fold) with the larger values being obtained in the more alcoholic media [7]. Similar results were found for the solvolyses in aqueous i-propanol and t-butanol [8]. However when the irradiation was introduced via the cup-horn (20 kHz — see Chapter 7) more substantial rate enhancements were obtained in ethanolic media. The general effects of sonication on this reaction are shown in Fig. 3.1 in which the ratio of first order rate constants with (k_{ult}) and without irradiation (k_{non}) is shown for various aqueous ethanol concentrations at different temperatures [9]

More detailed studies of the aqueous ethanol system led to the following main conclusions [10]:

(i) The effect of ultrasound increased with increased ethanol content and decreased temperature giving rate enhancements up to 20-fold at 10°C in 60% w/w.
(ii) At ethanol concentrations of 50 and 60% the actual rates of reaction under ultrasonic irradiation increased as the temperature was reduced from 20°C to 10°C (by factors of 1.4 and 2.1 respectively).
(iii) A maximum effect of ultrasound appeared to occur at a solvent composition of around 50% w/w at 25°C.

How then can we best explain the acceleration in reaction rate under ultrasonic irradiation? Perhaps the simplest way to understand how sonication can effect this solvolysis is to consider the energy profile for the reaction shown in Fig. 3.2.

In the slow, rate determining step of this reaction the energy barrier (E_a) is the difference in energy between the ground state of the substrate in solution and the first transition state of the reaction leading to the dissociation of the starting material into ions or ion pairs. Any process which serves to reduce this energy barrier will accelerate the reaction.

Under non-ultrasonic conditions, and at the same reaction temperature, a faster rate is obtained in a more aqueous solvent. The effect of ultrasound on the solvolysis might therefore be to liberate more 'free' water from the mixed solvent by disrupting the intermolecular hydrogen bonds. The result of such a disruption would be to increase solvation of the transition state by water and thus lower its energy. This would certainly help to explain effect (iii). Such a suggestion also gains support from the observations of Burton who, some forty years ago, investigated the hydrogen-bonded structure of aqueous ethanolic solvents using high frequency ultrasound and reported that a structural maximum existed for this binary solvent mixture at about

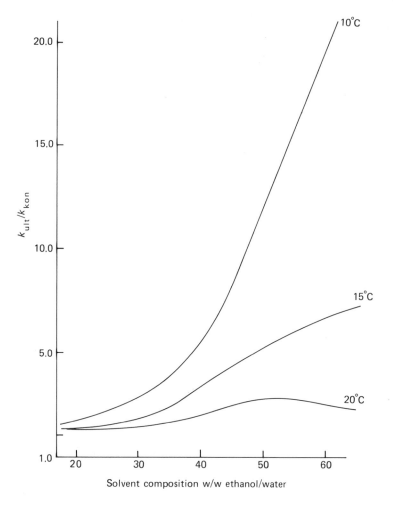

Fig. 3.1 — Effect of ultrasound on the solvolysis of 2-chloro-2-methylpropane in aqueous
ethanol mixtures.

0.3 mole fraction ethanol (*ca* 50% w/w) [11]. Thus at 25°C the maximum rate
enhancement might be expected to occur at a composition of 50% w/w (the position
of maximum structuredness) since ultrasonic irradiation should have its greatest
effect here.

What this rationale of simple solvent disruption does not seem to explain
however is the first two effects, the larger enhancements observed at lower temperatures which mask the maximum found at 25°C and the remarkable increase in
absolute value of k_{ult} as reaction temperature is lowered. To accommodate these
phenomena we must first bring in the increased sonochemical effect which is
expected as the solvent vapour pressure is lowered. Such reductions in vapour
pressure are achieved by decreasing the content of ethanol (vp 55.13 mmHg at 25°C)

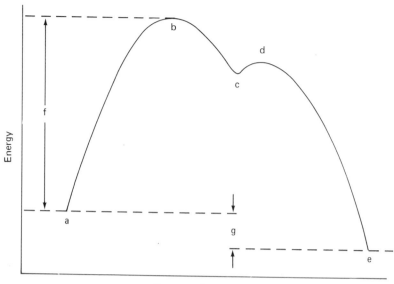

Fig. 3.2 — Energy profile for the solvolysis of 2-chloro-2-methylpropane in aqueous ethanol. (a) Starting material, (b) transition state (for breaking C–Cl bond), (c) intermediate carbocation (or ion pair), (d) transition state (for solvent attack on intermediate), (e) product, (f) activation energy, (g) enthalpy of reaction.

in water (vp 23.75 mmHg at 25°C) or by lowering the reaction temperature. Both factors are in operation as we consider the trends in k_{ult}/k_{non} in Fig. 3.1. In addition to this we should consider that irradiation could also be raising the ground state energy (i.e. pushing it along the energy profile towards ion pair formation). There are two ways in which this could occur (a) ultrasound could well disrupt the weak solvation of the low polarity ground state and thus increase its energy (such a disruption is much less likely in the almost ionic transition state) and/or (b) the increase in disruption of the bulk solvent, induced by ultrasound, should allow increased molecular mobility and thus higher collision frequencies. This should lead to a greater chance of energy transfer and hence provide more reactant molecules with enough energy to surmount the barrier. It is also possible that on the micro-scale the starting material itself could be subject to direct excitation due to the very high pressures and temperatures developed in the cavitation bubbles.

The effect of ultrasonic irradiation power on the chemical reactivity of this system has also been studied. Results show that for this hydrolysis reaction there is an optimum irradiation power for rate enhancement [10]. Such results suggest that for any ultrasonically enhanced reaction care must be taken in optimising conditions since sonochemical activation is not directly proportional to power input.

3.1.2 Using a direct immersion sonic horn

When we change the method of insonation from the bath, as used above, to a horn a major increase in ultrasonic power input to a reaction is possible. From typical bath powers of a few watts the direct introduction of ultrasound to a system via a horn will generate with ease sound energies in the range 20–50 watts cm^{-2}. The result is that some quite spectacular chemical reactions involving direct bond cleavage can be achieved.

Water itself can be decomposed with the formation of radical species and the production of oxygen gas and hydrogen peroxide (Scheme 3.1)

$$H_2O \rightarrow HO\cdot + H\cdot$$
$$H\cdot + O_2 \rightarrow HO_2\cdot$$
$$HO_2\cdot + HO_2\cdot \rightarrow H_2O_2 + O_2$$
$$HO\cdot + HO\cdot \rightarrow H_2O_2$$

Scheme 3.1

Any species dissolved in the water is clearly going to be subject to chemical reaction with these ultrasonically produced radicals and/or hydrogen peroxide. Thus if iodide ion is present in solution iodine will be liberated. Recent work using spin trapping e.s.r. techniques have given positive identification of the radical species sonically generated in water [12].

Although the first example of sonolysis in a non-aqueous solvent, the decolourisation of diphenylpicrylhydracyl (DPPH) radical in methanol, was reported in 1953 [13], it took some twenty years to realise that cavitation could successfully be supported in organic solvents. The lack of progress in this area was the result of a combination of two factors (i) the failure to observe, in organic media, certain sonochemical reactions which occurred in water and (ii) the knowledge that the addition of organic solutes suppressed sonochemically induced aqueous reactions. One of the problems encountered in such non-aqueous homogeneous sonochemistry is that solvents with high vapour pressures (e.g. ethers) are often used. To achieve effective cavitation in such solvents low temperatures are normally required. In recent times, with the advent of more powerful instrumentation a resurgence of interest in non-aqueous studies has occurred, most notably in the field of synthetic organic chemistry and polymer chemistry. The synthetic applications of non-aqueous sonochemistry will be discussed in this chapter but the two areas relating to polymers will be treated in a separate chapter.

Homogeneous non-aqueous sonochemistry is typified by the sonolysis of chloroform which has been studied using ultrasonic irradiation of frequency 300 kHz ($I = 3.5$ W cm^{-2}) to yield a large number of products amongst which are HCl, CCl_4 and C_2Cl_4 [14]. Decomposition was found to only occur in the presence of mono or diatomic gases, e.g. argon or nitrogen, but not with larger gas molecules like CO_2 (the underlying reasons for this, i.e. the dependence of the energy of cavitational collapse on γ the ratio of specific heats of the gas, has already been covered in Chapter 2). The precise mechanisms involved in the decomposition are complex but almost certainly involve the homolytic fission of chloroform to radicals and the formation of carbenoid intermediates as shown in Scheme 3.2.

Evidence for the generation of these reactive intermediates was obtained from a study of the effect of added cyclohexene on the sonication of chloroform. The presence of free radicals in the system was confirmed by the appearance of chlorocyclohexane as a product and by the increased rate of decomposition of $CHCl_3$ in the presence of cyclohexene. The increased decomposition rate is a consequence of the presence, in the cavitation bubble, of the alkene which scavenges or 'mops up' the $Cl\cdot$ radical as it is formed and prevents the regeneration of chloroform — i.e. the kinetic steps in Scheme 3.2 are *driven* from left to right. Carbene intermediates are implicated by the formation of tricyclic compounds such as (2) via dichlorocarbene addition to cyclohexene.

(2)

In 1983 Suslick reported the effects of high intensity (*ca* 100 W cm^{-2}, 20 kHz) irradiation of alkanes at 25°C under argon [15]. These conditions are, of course, well beyond those which would be produced in a reaction vessel immersed in an ultrasonic bath and indeed those normally used for sonochemistry with a probe. Under these extreme conditions the primary products were H_2, CH_4, C_2H_2 and shorter chain alk-1-enes. These results are not dissimilar from those produced by high temperature (>1200°C) alkane pyrolyses. The principal degradation process under ultrasonic irradiation was considered to be C–C bond fission with the production of radicals. By monitoring the decomposition of $Fe(CO)_5$ in different alkanes it was possible to demonstrate the inverse relationship between sonochemical effect (i.e. the energy of cavitational collapse) and solvent vapour pressure [16] (see Chapter 2).

The use of $Fe(CO)_5$ as a dosimeter in the above reaction was the result of previous studies on the decomposition of metal carbonyls in long-chain alkanes as solvents. Suslick studied the sonolysis of $Fe(CO)_5$ which produced $Fe_3(CO)_{12}$ together with finely divided iron (the proportion of each depending on the solvent vapour pressure). This is a significant result since it differs from both thermolysis (which yields finely divided iron) and ultraviolet photolysis (which yields $Fe_3(CO)_9$). The reaction is thought to proceed via the following mechanism (Scheme 3.3). It is interesting to note that this type of reaction cannot be performed in a cleaning bath [17] but this is almost certainly due to the fact that in an ultrasonic bath there is a greatly reduced intensity of sonication compared with that achievable with a direct immersion horn.

This mechanism was supported by evidence obtained from the sonolysis of any one of the three individual iron carbonyls $Fe(CO)_5$, $Fe_2(CO)_9$ and $Fe_3(CO)_{12}$ in the presence of pent-1-ene [18]. In each case the pent-1-ene isomerised to give the same *cis*- to *trans*- isomer ratio of pent-2-ene indicating that the same catalytic species is generated from each carbonyl. For the same system, ligand substitution by phosphines or phosphites occurred readily to give $Fe(CO)_{5-n}L$ ($n = 1, 2, 3$). The rates of these reactions were independent of ligand concentration which lends further

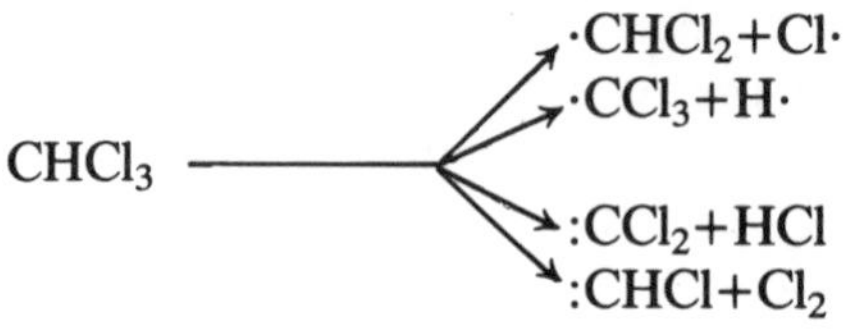

$$\text{CHCl}_3 \quad
\begin{cases}
\nearrow \cdot\text{CHCl}_2 + \text{Cl}\cdot \\
\nearrow \cdot\text{CCl}_3 + \text{H}\cdot \\
\searrow :\text{CCl}_2 + \text{HCl} \\
\searrow :\text{CHCl} + \text{Cl}_2
\end{cases}$$

Scheme 3.2

$$\text{Fe(CO)}_5 \rightarrow \text{Fe(CO)}_{5-n} + n\text{CO}$$
$$\text{Fe(CO)}_5 + \text{Fe(CO)}_3 \rightarrow \text{Fe}_2(\text{CO})_8$$
$$2\text{Fe(CO)}_4 \rightarrow \text{Fe}_2(\text{CO})_8$$
$$\text{Fe(CO)}_5 + \text{Fe}_2(\text{CO})_8 \rightarrow \text{Fe}_3(\text{CO})_{12} + \text{CO}$$

Scheme 3.3

support to a mechanism involving the ultrasonic generation of a highly reactive co-ordinatively unsaturated intermediate. Similar ligand exchange reactions have been observed with $\text{Mn}_2(\text{CO})_{12}$ and $\text{Re}_2(\text{CO})_{10}$ [19].

3.2 HETEROGENEOUS SONOCHEMISTRY

The majority of reactions reported to be enhanced by ultrasonic irradiation are of the heterogeneous type and are either liquid/liquid or liquid/solid. In general there are two methods of introducing the ultrasound — either with an ultrasonic cleaning bath or a probe system (see above). Undoubtedly the majority of sonochemical studies to date have been performed with a cleaning bath — by far the simplest technique but not necessarily the most efficient.

3.2.1 Heterogeneous reactions involving a metal as a reagent

Using a cleaning bath
Ultrasound is widely used in organometallic chemistry for the preparation of organometallics from reactive metals. Here the mode of action of ultrasound is primarily an erosion phenomenon involving the removal of oxide layers and impurities together with pitting of the metal surface (which increases the possible reaction area). Surfaces thus treated contain an increased number of dislocations which are widely considered to be the active sites in catalysis.

The mechanism of the effect of ultrasound on surfaces is two-fold: (i) *Acoustic streaming* (which aids mass transport) is the movement of the liquid induced by the sonic wave which can be considered to be simply the conversion of sound to kinetic energy and is not a cavitational effect. (ii) *Pitting of the metal surface*, on the other hand, is a direct result of transient bubble collapse near the surface. The presence of an interface causes the formation of asymmetric cavities. During collapse this irregularity becomes pronounced and a jet of liquid is directed towards the surface — this phenomenon has actually been recorded using high-speed microphotography.

Along with the shock wave associated with cavitation this causes localised deformation of the surface together with fragmentation and reduction in particle size.

In 1980 Luche published the first of a series of papers dealing with ultrasound promoted reactions of organometallics of use in organic synthesis [20]. He reported the direct, *in situ*, formation of alkyl and aryl lithiums by the reaction of an organic halide with lithium wire (or lithium — 2% sodium sand) in ether immersed in an ultrasonic bath (50 kHz). The technique avoids the use of activating reagents (e.g. I_2 or iodomethane) and affords a significant amelioration of the reaction both by increasing reactivity and removing the induction period which is often involved. This methodology finds a particularly notable application in the Barbier reaction — the one-step coupling of an organic halide with a carbonyl compound in the presence of magnesium or lithium metal. Significant improvements both in yields and simplification of experimental techniques over conventional methods are obtained when the reactions are carried out in an ultrasonic cleaning bath (50 kHz). A particular advantage of this method is that the syntheses are largely free from side reactions such as reduction and enolisation which are common in conventional methodology. Thus allylic alcohol (**3**) has been isolated in 96% yield from the reaction between hexanal and 2-bromopropene (eqn 3.4). Even benzyl halides gave yields in excess of 95% and very little Würtz coupling was observed, which often predominates with non-ultrasonic methods. A further advantage is that such reactions can be performed in damp, technical grade tetrahydrofuran, a potential boon for large-scale industrial operations.

$$\text{(3.4)}$$

(3)

Today one of the most common chemical applications of ultrasound is the initiation of a reluctant Grignard reaction. The quantitative effects of ultrasound on the induction times for the formation of a Grignard reagent in various grades of ether is given in Table 3.2 [21].

The term 'crushed' in Table 3.2 refers to the methodology commonly employed to activate magnesium — mechanical crushing of the metal to expose fresh surface to the reactants. In each of the three solvents there was no significant difference in yield of organometal with or without sonication, the figures being 65, 55 and 55%. Significantly, prior sonication of the metal in ether had no effect on the induction time under non-ultrasonic conditions. This clearly eliminates simple surface cleaning as the source of the effect and it was suggested that during sonication adsorbed water was removed from the metal surface, kept clear while irradiation continued, but was re-adsorbed on switching off the bath.

The enhancements which have been obtained in the preparation of organo lithium reagents have already been referred to [20]. For example *n*-propyl, *n*-butyl and phenyl lithium have been prepared in >90% yield by reaction of the appropriate bromide with Li wire (or Li-2% Na sand) in ethereal solvents (eqn 3.5). Although the reactions with primary bromoalkanes commenced immediately, secondary and

Table 3.2 — The preparation of butan-2-yl magnesium bromide in ether in an ultrasonic bath (50 kHz)

$$CH_3CH_2CHBrCH_3 + Mg \rightarrow CH_3CH_2CH(MgBr)CH_3$$

Type of diethyl ether used	Method	Induction time
Pure, dried (0.01% water)	Stirred	6–7 min
(0.01% ethanol)	Sonicated	less than 10 s
Reagent grade (0.5% water)	Stirred	2–3 h (*crushed*)
(2.0% ethanol)	Sonicated	3–4 min
50% saturated (0.01%) ethanol	Stirred	1–3 h (*crushed*)
	Sonicated	6–8 min

tertiary alkyl bromides required longer periods of sonication. Similarly, rate enhancements were reported for the formation of sodium naphthalene in commercial undried THF [22].

$$R\,Br\ +\ Li\ \longrightarrow\ R\,Li\ +\ Li\,Br \qquad (3.5)$$

A very reactive form of finely divided metals are the so-called Rieke powders [23]. These are produced as fine powders by chemical precipitation during the reduction of various metal halides with potassium metal in refluxing tetrahydrofuran. Obviously this is a potentially hazardous laboratory procedure and ultrasound has provided an alternative method of preparation of these extremely valuable reagents [24]. The sonochemical technique involves the reduction of metal halides with lithium in THF at room temperature in a cleaning bath (50 kHz) and gives rise to metal powders which have reactivities comparable to those of Rieke powders. Thus powders of Zn, Mg, Cr, Cu, Ni, Pd, Co and Pb were obtained in less than 40 minutes by this ultrasonic method compared with reaction times of 8 hours using the experimentally more difficult Rieke method (Table 3.3).

These sonically prepared Rieke powders show enhanced reactivity in organic synthesis involving metals, e.g. the Reformatsky and Ullmann coupling reactions (see later).

A highly active form of metallic magnesium is formed when commercial magnesium powder in THF is subjected to low intensity ultrasound in the presence of anthracene [25]. The anthracene forms an electron transfer complex with magnesium and this effectively acts as a phase transfer agent (Scheme 3.4). The magnesium

Table 3 — The generation of Rieke powders

$$MX_n + nA \rightarrow M^* + nAX$$
$$X = Cl, Br, I$$
$$A = Li, Na, K$$

	Method	Time
$CuI_2 + K$	Rewflux in THF	8 h
$CuBr_2 + Li$	Ultrasound/room temperature/THF	less than 40 min
$NiCl_2 + Li$ (Powder)	Stir/room temperature	14 h
$NiCl_2 + Li$ (Powder)	Ultrasonic bath	less than 40 min

produced in this way is an excellent reducing agent for metal salts and when the reduction is carried out in the presence of Lewis base ligands it is a useful route to organo-transition metal complexes for example, η^5-cyclopentadienyl complexes Cp_2M (M=V, Fe, Co), η^3-allyl complexes (M=Co, Ni), alkene complexes (M=Ni, Pd, Pt, Mo) and phosphine complexes (M = Pd, Pt). The magnesium anthracene complex is also a convenient route to allyl Grignard reagents at temperatures as low as $-35°C$. This eliminates the coupling of allyl magnesium halides with the starting alkyl halide which is a common side reaction with conventional methods (eqn 3.6) [26].

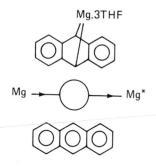

Scheme 3.4

MgBr + Br → (3.6)

Sonication (35 kHz cleaning bath) is a simple and convenient method for preparing colloidal alkali metals [27]. For example the blue colour typical of colloidal potassium is produced in a few minutes when the metal is insonated in toluene or xylene. By way of contrast it is possible to disperse metallic sodium in xylene (b.p. 138°C) but not in toluene (b.p. 110°C). This is another example of the more powerful cavitation in the higher boiling solvents being an important factor. These colloidal systems are particularly useful in condensations of the Dieckmann type, in the generation of ylide intermediates in Wittig reactions and in the reductive cleavage of cyclic sulphones (Scheme 3.5) [28].

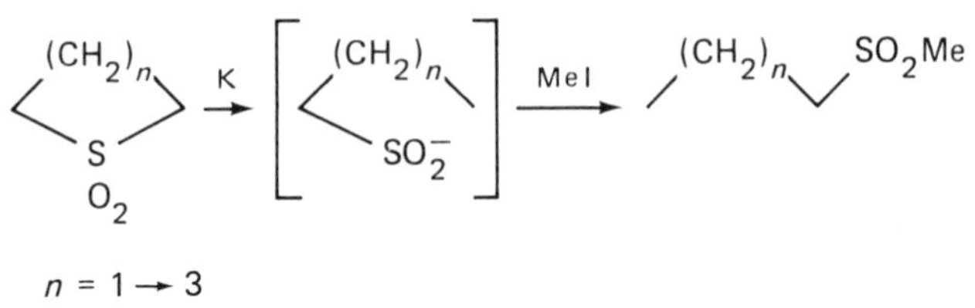

$$n = 1 \rightarrow 3$$

Scheme 3.5

The Bouveault reaction is the preparation of an aldehyde by a one pot reaction between an organic halide and lithium metal in dimethylformamide. Ultrasound has been found to markedly enhance this reaction when it is performed in tetrahydrofuran [29]. Use of an ultrasonic bath (40 kHz) at 10–20°C affords short reaction times of between 5 and 15 minutes and generates yields in the range 70–88%. Using this methodology the conversion of 1-bromobutane to pentanal (88%) can be achieved in only 5 minutes. This must be contrasted with the yield of less than 10% which is obtained under the normal stirred conditions in the same time period. This result confirms that the effect of irradiation goes beyond mere agitation (eqn 3.7).

Low intensity ultrasound also facilitates the Reformatsky reaction of α-haloesters with aldehydes and ketones (eqn 3.8) [30]. The ultrasonic method gave ethyl 2-hydroxyphenylacetate (4, 98%) in 5 min for R = C_6H_5 and R' = H compared with 98% obtained in 1 hour using activated zinc powders, 95% in 5 hours using the $(MeO)_3B/THF$ solvent system and only 61% after 12 hours using the conventional methodology. Similarly the reaction between ethyl bromoacetate and cyclopentanone generated (5, 98%) in 30 minutes at 25°C compared with 80% in 12 hours at 80°C by conventional methods. It is interesting to note that these improved yields were obtained using dioxane as solvent. With the more conventional, but lower boiling, Reformatsky solvents — benzene or ether — there was no improvement using ultrasound. This lends strong support to the importance of the choice of solvent when embarking upon sonochemistry research.

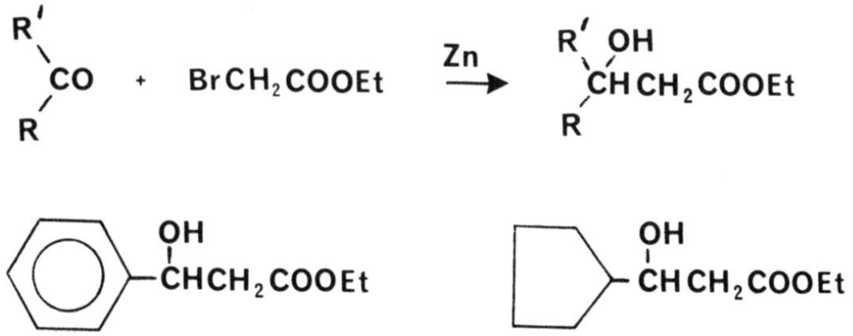
R'
CO
R
BrCH2COOEt
Zn
R' OH
CHCH2COOEt
R
OH
CHCH2COOEt
OH
CHCH2COOEt

(3.8)

(4) (5)

Ultrasonic irradiation of a mixture of zinc and α,α'-dibromo-*ortho*-xylene in dioxane results in dehalogenation and the generation of a xylylene intermediate (**6**) which readily adds to any dienophiles present in the reaction mixture, e.g. with maleic anhydride or methyl propenoate to afford high yields of (**7**) and (**8**) respectively [31]. In the absence of dienophile the product is mainly polymer with a trace of (**9**) (Scheme 3.6). The work was performed in a bath (50/60 Hz) at 25°C on 10 mmol scale using 23 mmol zinc under N_2. There was no reaction in the absence of ultrasound.

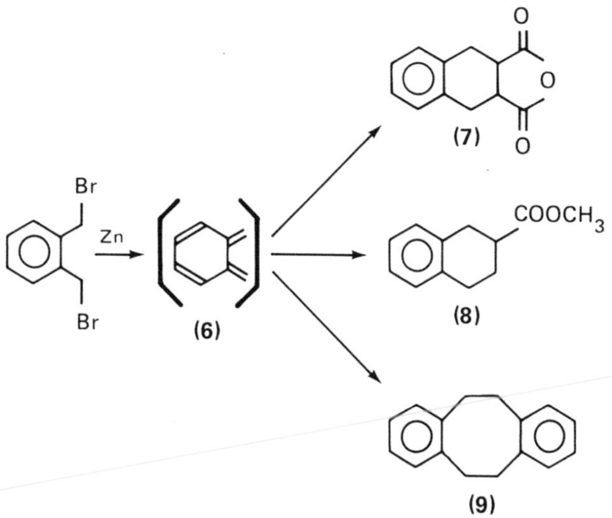
Br
Zn
Br
(6)
O
O
(7)
O
COOCH3
(8)
(9)

Scheme 3.6

The zinc promoted cycloaddition of α,α'-dibromoketones to 1,3-dienes is enhanced by ultrasound to such an extent that compounds such as the highly hindered bicyclo[3.2.1]oct-6-en-3-ones (**10**) become easily accessible (eqn 3.9) [32].

Reactions of this type, in the absence of ultrasound, give only low to moderate yields and require much longer reaction times of the order of 24 h.

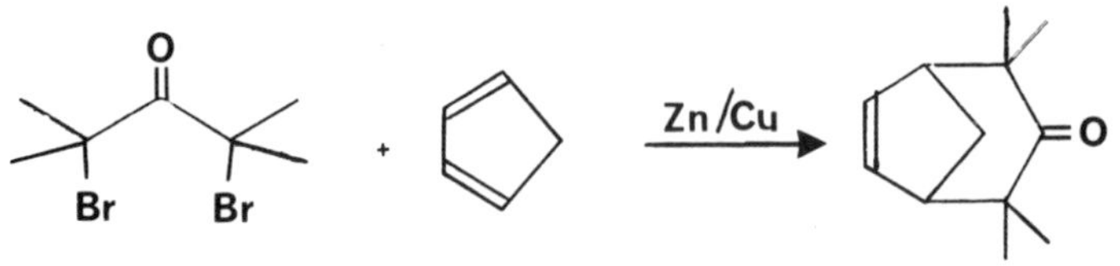

(3.9)

(10)

The (2+2) cycloaddition of dichloroketene to alkenes is also improved by ultrasound (eqn 3.10) [33]. The ultrasonic method shows all of the advantages normally associated with the technique i.e. short reaction times, the formation of good yields at ambient (rather than elevated) temperatures and the use of ordinary (commercial) zinc dust rather than the more exotic zinc/copper couple.

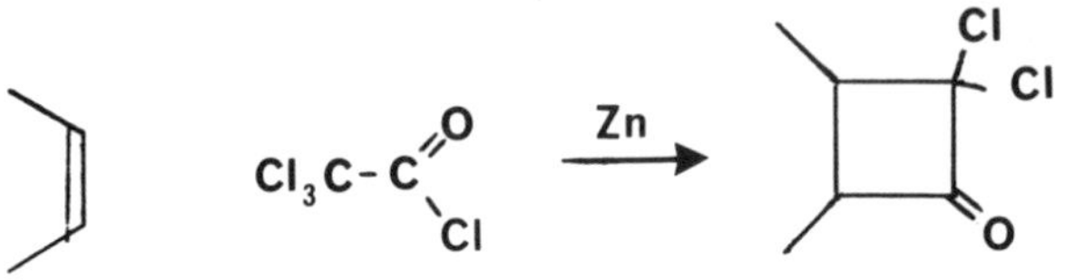

(3.10)

A convenient method for the conversion of aldehydes (RCHO) to alkenes $(RCH=CH_2)$, known as methylenation, involves the reaction of a zinc/copper couple with diiodomethane in the presence of the carbonyl compound dissolved in tetra-hydrofuran. The reaction first generates an organometallic intermediate (ICH_2ZnI) which then reacts with the carbonyl compound. The conversion of benzaldehyde to styrene using this conventional methodology required a reaction time of 6 hours at 40°C. When the reaction was sonicated however comparable yields of around 70% were obtained in the production of the corresponding styrenes from benzaldehyde, 4-chlorobenzaldehyde and 4-methylbenzaldehyde after irradiation times of 20, 15 and 120 minutes respectively at room temperature (eqn 3.11) [34]. Somewhat poorer yields are obtained from the methylenation reactions of ketones.

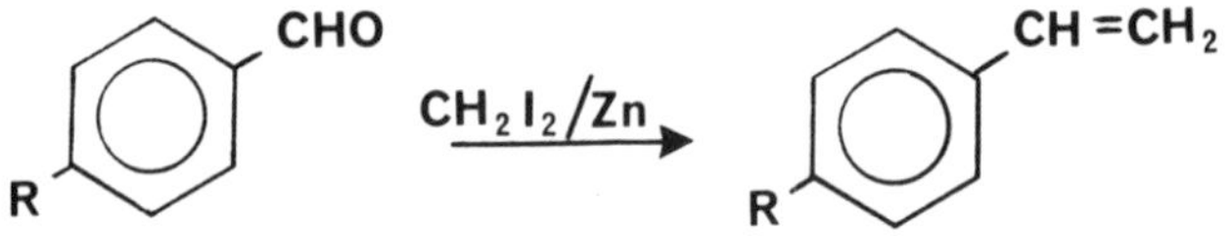

(3.11)

When the Zn/CH_2I_2 reaction is applied to an alkene the result is the formation of of a cyclopropane and this is generally known as the Simmons–Smith reaction. Unfortunately, while the reaction itself is quite useful, on a preparative scale it suffers from several drawbacks one of the major ones being the sudden exotherm which occurs after an unpredictable induction period. In 1982 Repic described a modification of the Simmons–Smith reaction using sonochemically activated zinc which eliminated the sudden exotherm normally associated with the reaction. Using ultrasound it was found that (**11**) could be produced in 91% yield compared with 51% by the normal route (eqn 3.12) [35]. The normal method for enhancing this reaction relied upon activation of the zinc metal by using it in the form of a zinc-silver or zinc-copper couple and/or using iodine or lithium in conjunction with the metal. In the sonochemical procedure no special activation of the zinc was required and in fact equally good and reproducible yields were obtained using zinc dust or even the metal in the form of mossy rods or foil. This methodology has now been successfully scaled up to run in a 22 dm^3 vessel immersed in a 50-gallon bath (Fig. 3.3). For the scale-up

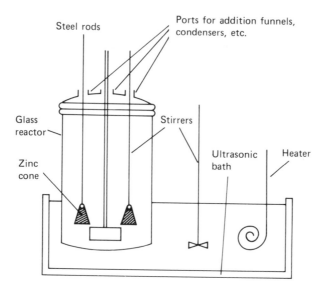

Fig. 3.3 — Apparatus for large-scale cyclopropanation.

the zinc was cast in two 800 g lumps (using 125 cm^3 conical flasks as moulds). The reagents were methyl oleate (0.6 kg), diiodomethane (1.3 dm^3), dimethyloxyethane (2.7 dm^3). Under nitrogen at 100°C the reaction yielded (**11**) 0.5 kg (82%) after 2.25 h [36].

$$CH_3(CH_2)_7 \overset{}{\diagup}\overline{}\overset{}{\diagdown}(CH_2)_7CH_3 \xrightarrow{\ Zn\ } CH_3(CH_2)_7 \overset{\triangle}{\diagup}\diagdown(CH_2)_7CH_3 \qquad (3.12)$$

(11)

The method has several advantages over the normal method of cyclopropanation which involves several experimental problems:

(1) There is a reduction in foaming (normally associated with ethene and cyclopro-
    pane formation);
(2) the exotherm is more evenly distributed (only a small clean area of metal is
    available throughout the reaction);
(3) the reaction can be controlled by removing the lump of metal from the reaction
    (this is not unlike the use of fuel rods in a nuclear reactor);
(4) the residual metal can be removed from the reaction as a lump.

Methods for the introduction of the perfluoroalkyl group R_f into organic
compounds are not as plentiful as those for the introduction of simple alkyl groups.
The standard organometallic routes for perfluoroalkyl substitution are not generally
available due to the low stability or low reactivity of the perfluoroalkylmetal
intermediates, thus both R_fMgX and R_fLi readily decompose into perfluoroalkenes.
Ultrasound has now provided a solution to this problem and a range of synthetic
routes have been developed, each starting from a perfluoroalkane and employing
powdered zinc (Scheme 3.7) [37].

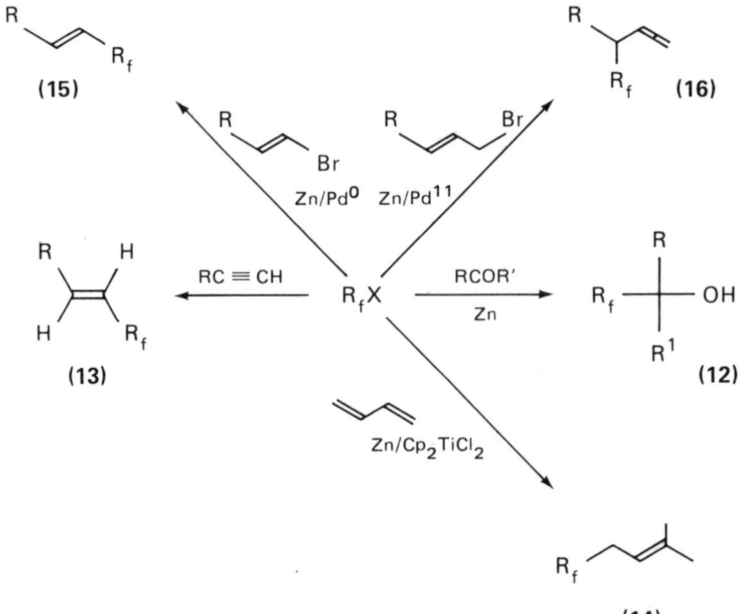

Scheme 3.7

The synthesis of alcohol (12) can be achieved by a form of Barbier reaction in
which a carbonyl compound, in DMF as solvent, reacts with perfluoroalkylzinc
which is sonochemically generated *in situ*. Hydroperfluoroalkylation can be

achieved by the direct reaction of alkynes with perfluoroalkylcuprates, again generated *in situ* from the reaction of perfluoroalkyl halides with zinc and copper(I) iodide in THF. Good yields of fluoroalkyl substituted alkenes (13) are obtained and the reaction is regiospecific (but not stereospecific). Similar results are obtained in the preparation of (14) by the perfluoroalkylation of dienes catalysed by Cp_2TiCl_2 (Cp = cyclopentadienyl). High yield cross-coupling reactions of vinyl halides with perfluoroalkylzinc reagents to (15) are catalysed by Pd(0) [Pd(PPh$_3$)$_4$] while similar reactions of allyl halides to (16) require Pd(11) [Pd(OAc)$_2$]. The same group have also reported a synthesis of chiral ketones (eqn 3.13) via the optically active enamine (17). Sonication of zinc powder and Cp_2TiCl_2 in THF produces *bis*-[η^5-cyclopentadienyl]titanium(11) which is believed to be the catalytic species involved in the degradation of (17) to the target ketone in optical purities of between 60 and 70%.

(17)   (3.13)

Asymmetric induction of between 30 and 66% has been reported for the addition of perfluoroalkyl iodides to chiral arene/chromiumtricarbonyl complexes using ultrasonically dispersed zinc at room temperature (eqn 3.14) [38]. Photochemical decomposition of the organometal intermediate affords a chiral alcohol product. The reaction is carried out in DMF as solvent and high overall yields are reported e.g. 80% (R=Et). Only a small excess of the perfluoroalkyliodide is required and the conversion is complete in under one hour.

(3.14)

The alkenyl epoxides (18, R^1 = Me, R^2 = H, Ph, 1-hexanyl) are smoothly converted to the corresponding ferrilactone complexes (19) on reaction with $Fe_2(CO)_9$ suspended in THF and sonicated at room temperature [17]. Such complexes undergo several synthetically useful transformations (Scheme 3.8) including oxidation with Ce(IV) as a route to β-lactone natural products or β-lactam antibiotics and reaction with CO to afford δ-lactones.

Using a Probe

Two types of reaction can be chosen to illustrate the way in which a move to irradiation from a horn can be used to obtain more effective reaction enhancement

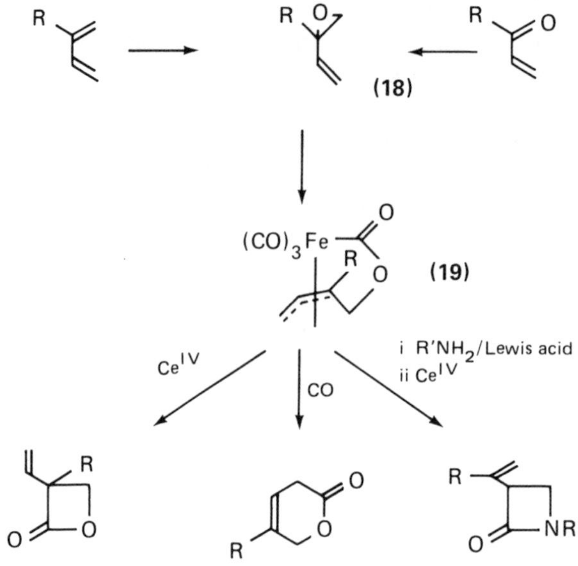

Scheme 3.8

than that provided by the ultrasonic bath. Both types of reaction — the preparation of organozinc reagents and the use of copper bronze in an Ullmann coupling reaction — were first explored in baths.

Diaryl zincs (**20**) can be prepared by sonication of a mixture of aryl bromide, zinc bromide and lithium metal in dry ether or tetrahydrofuran in a cleaning bath [39]. Generation of the diaryl zinc takes 30–45 minutes after which the cleaning bath can be turned off and the substrate added to the mechanically stirred solution together with nickel acetylacetonate (Ni(acac)$_2$) which acts as a catalyst. In this way conjugate addition can be achieved in relatively high yield. For example (**20**, R = 2-CH$_3$C$_6$H$_4$) adds to 3-methylcyclopentenone (72%) compared with only a 20% yield under normal conditions (eqn 3.15). This procedure proved to be less effective for dialkyl zinc reagents but a modification has led to a more generalised method in which an ultrasonic probe has replaced the cleaning bath [40]. Using toluene:tetrahydrofuran (85:15) as solvent, dialkyl or diaryl zincs can be generated in 10–15 minutes at 0°C. The specially designed apparatus used for this preparation is shown (Fig. 3.4). The organolithium is first formed by direct sonication of the metal and this is transformed immediately into the organozinc. Subsequent addition of an enone, in the absence of irradiation, gives rapid reaction at room temperature in the presence of Ni(acac)$_2$ (Scheme 3.9). This approach has proved to be a particularly useful alternative to the use of either lithium dialkyl copper reagents (LiCuR$_2$) or copper catalysed Grignard additions for conjugate addition. Thus while (+)-cuparenone (**21**) cannot be prepared by these conventional techniques from 4,4-dimethyl-3-*p*-tolylcyclopentenone, the conversion can be achieved smoothly and in 84% yield via sonochemically prepared dimethyl zinc. Similar conjugate additions have also been achieved with cyclohexenone (**22**, 88%) and unsaturated aldehydes (e.g. **23**, 69%).

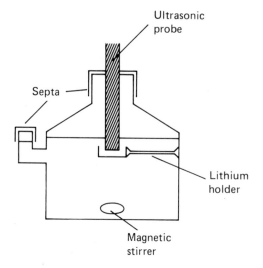

Fig. 3.4

Scheme 3.9

$$ZnR_2 \quad + \qquad\qquad\qquad \longrightarrow \qquad\qquad\qquad\qquad\qquad (3.15)$$

(20)

An Ullmann coupling reaction is the linking of two aryl rings by reaction of the corresponding haloarenes with an excess of copper metal in a high boiling solvent. When the arene is activated by the presence of an *ortho* nitro group the reaction can be performed under milder conditions. For example 2-iodonitrobenzene can be converted to the corresponding biphenyl (**24**) in 48 h at 60°C using a ten-fold excess of copper bronze (eqn 3.16) [14]. For such activated aryl halides, dimethylformamide (DMF) is a recommended solvent for the coupling reaction and commercial copper bronze or freshly precipitated copper obtained by reduction of a copper salt have been claimed to give satisfactory results. Pretreatment of the copper with EDTA or iodine to remove surface impurities is claimed to give more consistent results. It was the importance of clean copper surfaces in these reactions together with the well known cleansing action of ultrasound which led us to investigate the effect of ultrasonic irradiation on this Ullmann coupling reaction. Initial studies were carried out using an ultrasonic laboratory cleaning bath operating at 45 kHz.

$$(3.16)$$

(24)

Although our results showed a rate enhancement of 60% over the control experiment the total conversion was only 21% over two hours. We next investigated the use of a 1.5 mm microtip probe driven by a heat systems W225 sonicator (20 kHz) operating at 60% power as the source of ultrasound. The results using a 1:1 mole ratio of copper (15 mmol) to 2-iodonitrobenzene (15 mmol) in DMF (40 cm^3) at 65°C gave 48% (**24**) in 2 h. Investigations into the effect of copper to halide ratio revealed that a 4:1 ratio is optimum yielding 80% product in 1.5 h. During these studies it was observed that the average particle size of the copper, as determined by laser light scattering, fell from 87 μm to 25 μm after 1 hour of sonication in DMF. This factor alone however was shown to be insufficient to explain the large (fifty-fold) enhancement in reactivity produced by ultrasonic irradiation. These studies suggested several advantages in using ultrasound:

(i) It cleaned the metal surface;
(ii) It reduced the particle size of the metal;

(iii) It increased the rate, suggesting that sonication assisted in breaking down intermediates and/or assisted in the desorption of products;

(iv) It prevented the adsorption of copper on the walls of reaction vessels which is a common problem with conventional methodology.

The use of ultrasound has also been found to dramatically reduce the temperatures and pressures required for the preparation of early transition metal carbonyl anions from the direct reaction of the corresponding metal chlorides with carbon monoxide. Prolonged irradiation (15 h at 10°C using a 20 kHz probe) of the appropriate halide together with sodium sand in tetrahydrofuran through which carbon monoxide is bubbled leads to approximately 50% yields of $W_2(CO)_{10}^{2-}$, $Mo_2(CO)_{10}^{2-}$ and $Nb(CO)_6^{2-}$ [42]. Using this technique $V(CO)_6^-$ (35%) can be prepared at 4.4 atm pressure whereas normal conditions require 200 atm and 160°C (eqn 3.17). The suggested mechanism for such carbonylations involves the direct trapping by carbon monoxide of a reactive metal species produced during the reduction of the metal halide.

$$VCl_3(THF)_3 \xrightarrow{\text{CO/Na/THF}} V(CO)_6^- \tag{3.17}$$

3.2.2 Heterogeneous reactions involving non-metallic reagents
Using a bath

The Cannizzaro reaction under heterogeneous conditions (solid–liquid) using barium hydroxide as base is greatly accelerated by low intensity ultrasound (cleaning bath). Yields of 100% of the disproportionation products are obtained after 10 min sonication of benzaldehyde whereas no reaction is observed during this period in the absence of ultrasound (eqn 3.18) [43].

$$\text{PhCHO} \xrightarrow[\text{EtOH}]{Ba(OH)_2} \text{PhCOOH} + \text{PhCH}_2\text{OH} \tag{3.18}$$

Low intensity ultrasound has a marked accelerating effect on hydroborations which are traditionally very slow, the effect being particularly marked in the heterogeneous systems. For example, the preparation of tricyclohexylborane by hydroboration with $BH_3 . SMe_2$ in THF traditionally requires 24 h at 25°C, however, with irradiation, in an ultrasonic bath (50 kHz, 150 W) the reaction is complete in 1 h (eqn 3.19) [44]. Similarly the hydroboration of *trans*-hex-3-ene with $HBBr_2 . SMe_2$ (eqn 3.20) which normally requires 12 h at 25°C is completed in only 2 h in an ultrasonic bath.

$$\text{(3.19)}$$

$$\text{(3.20)}$$

Ultrasound, when used in a solid/liquid reacting system, shows promise as a method of avoiding the use of phase transfer catalysts. This has been found to be particularly relevant in the generation of dichlorocarbene by the direct reaction between powdered sodium hydroxide and chloroform [45]. The reported procedure is both simple and efficient in that irradiation of a stirred mixture of powdered NaOH in chloroform containing an alkene generates high yields of the corresponding dichlorocyclopropanes in 4 h at 40°C. The results make it quite clear that both mechanical stirring and ultrasonic irradiation are necessary for high reactivity thus whereas styrene is cyclopropanated in 96% yield in 1 h when a combination of both sonication and stirring is used, the yield is much reduced to 38% in 20 h with sonication alone and to 31% in 16 h with only mechanical stirring (eqn 3.21).

$$\text{(3.21)}$$

In the ultrasonic preparation of aromatic acyl cyanides (25, R=H, 2-Me, 3-Me, 4-Me, 4-MeO) (eqn 3.22) by the reaction of the corresponding acyl chlorides with solid KCN in acetonitrile good yields of products have been achieved (70–85%) even in the absence of phase transfer catalysts [46]. Under non-ultrasonic conditions this reaction is facilitated by the presence of traces of water but proceeds very slowly, whereas the ultrasonic reactions are not significantly affected by water. A possible rationale for this may be that the role of the water is to attack the crystal lattice to reveal sites which may well be more easily exposed by ultrasound.

$$R - \underset{\underset{O}{\|}}{C} - Cl \; + \; KCN \longrightarrow R - \underset{\underset{O}{\|}}{C} - CN$$

$$\text{(3.22)}$$

(25)

Other displacement reactions which have been found to benefit from ultrasonic irradiation are the generation of azides from activated primary halides, e.g. allyl azide (91%) from allyl chloride and aqueous sodium azide (eqn 3.23) [47], and the quantitative preparation of isotopically labelled 17-I-heptadecanoic acid (eqn 3.24) from the corresponding bromoacid in MEK in the presence of Na_2SO_3 [48]. The method employed in the latter case involves sonication at 100°C of a mixture of bromoacid, saturated thiosulphate and ^{123}I-labelled iodine in butan-2-one.

$$\text{CH}_2\!\!=\!\!\text{CH}\!\!-\!\!\text{CH}_2\text{Cl} \ + \ \text{Na N}_3 \ \longrightarrow \ \text{CH}_2\!\!=\!\!\text{CH}\!\!-\!\!\text{CH}_2\text{N}_3 \qquad (3.23)$$

$$\text{Br}-(\text{CH}_2)_{16}\text{COOH} \ + \ \overset{*}{\text{I}}_2 \ \xrightarrow{\ \text{Na}_2\text{S}_2\text{O}_3\ } \ \overset{*}{\text{I}}-(\text{CH}_2)_{16}\text{COOH} \qquad (3.24)$$

In 1982 Boudjouk published studies of the reduction of aromatic halides with lithium aluminium hydride. Under normal conditions deactivated halides are not reduced significantly by this reagent, indeed the reduction of bromobenzene to benzene (52%) in tetrahydrofuran requires a four-fold excess of $LiAlH_4$ and a reaction time of 6 hours. In the particular cases of the bromobenzenes (**26**, R=CH$_3$, OCH$_3$) the yields of the respective hydrocarbon after 24 h were only 20% (tetra-hydrofuran, 25°C) and 35% (dimethylformamide, 100°C) (eqn 3.25). When these reductions were carried out in an ultrasonic bath (50/60 kHz) at 35°C in dimethox-yethane the yields were dramatically improved to 97 and 70% in only 6 hours [49].

$$\text{(26)} \quad \text{C}_6\text{H}_4(\text{R})\text{Br} \ + \ \text{LiAlH}_4 \ \longrightarrow \ \text{C}_6\text{H}_4(\text{R})\text{H} \qquad (3.25)$$

$$(\text{CH}_3)_3\text{X Cl} \ + \ \text{LiAlH}_4 \ \longrightarrow \ (\text{CH}_3)_3\text{X H} \qquad (3.26)$$

$$\text{(27)}$$

Hydrides of the group IVB elements (e.g. **27**, X=Si,Sn) have been prepared in high yields by the heterogeneous reduction of the corresponding halo, alkoxy and amino derivatives using $LiAlH_4$ suspended in non-polar (hydrocarbon) solvents under ultrasonic irradiation [50]. Without sonication these reductions did not take place at all. Using a three-fold excess of the reducing agent (**27**, X=Si) (80%) was produced in 3 h at 40°C and (**20**, X=Sn) (>95%) in 2.5 h at 25°C.

The oxidation of alcohols by solid potassium permanganate in hexane and benzene is also significantly enhanced by irradiation in an ultrasonic bath. Compari-sons with simple mechanical agitation reveal increases in yield at 50°C for the oxidation of octan-2-ol to octan-2-one of from 2.6 to 92.8% (5 h) and for 3-phenylprop-2-en-1-ol from 4.5 to 82.8% (3 h) [51].

Using a probe

One of the earliest reports of the acceleration of a biphasic reaction was the hydrolysis of carboxylic acid esters (**28**, Ar = phenyl,2,4,6-trimethylphenyl,2,4-and 3,5-dimethylphenyl) in aqueous sodium hydroxide (eqn 3.27) [52]. This work was reported in 1979 and, unusually, was performed using an ultrasonic probe (20 kHz) rather than the (then) more commonly used cleaning bath. With this technique the reaction time for the hydrolysis of methyl benzoate (**28**, Ar=phenyl) was reduced from 1.5 h at reflux to just 10 min at 25°C. The authors discounted bulk heating effects as the cause of the rate enhancement.

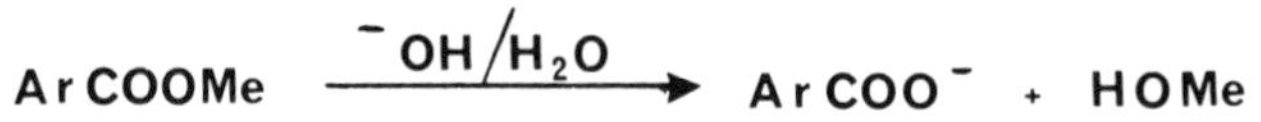

$$ArCOOMe \xrightarrow{\ ^{-}OH/H_2O\ } ArCOO^{-} + HOMe \qquad (3.27)$$

(28)

The *O*-alkylation of hindered phenols are normally sluggish reactions and attempts to improve reactivity have received considerable attention from synthetic organic chemists. Ultrasonic irradiation has been used to greatly enhance the rate of alkylation of 2,6-dimethyl phenol with several different primary halides under heterogeneous conditions using K_2CO_3 in *N*-methylpyrrolidinone (NMP) (eqn 3.28) [53].

Table 3.4 — Yield of ether from the alkylation of 2,6-dimethylphenol

Halocompound	Yield % (glc) after 1.5 h	
	Sonicated	Normal
Iodomethane	90	45 (60 after 4 h)
1-iodopropane	95	33 (45 after 4 h)
3-bromoprop-1-ene	100	38

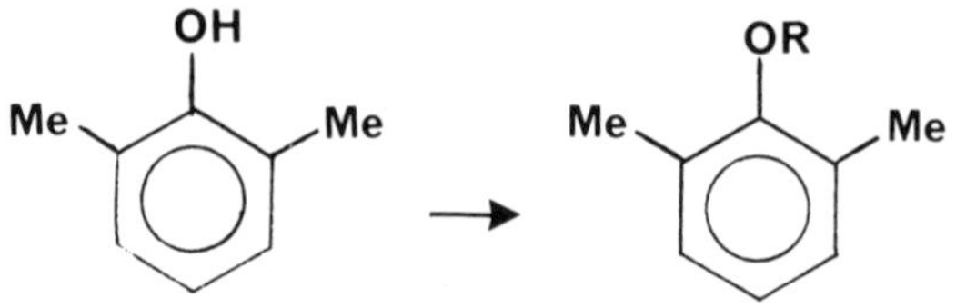

(3.28)

The results in Table 3.4 show that the most obvious benefit of ultrasonic irradiation on these *O*-alkylation reactions is the substantial acceleration in rate giving almost complete alkylation in about 1.5 h for all three systems. Table 3.4 also reveals a characteristic of such alkylations namely that under normal, stirred, conditions there is a distinct fall-off in reactivity as the reaction proceeds. Since it is known that sonication is capable of reducing particle size in reactions involving powders, the possibility that simple particle size reduction (i.e. increase in surface area) of the potassium carbonate might explain the sonochemical enhancement was investigated. This was accomplished by first suspending K_2CO_3 in NMP and subjecting it to sonication for 1 h. During this period the initial agglomerates of powder, some 100–300 μm in diameter, were broken down to particles with a fairly even size distribution of around 3–5 μm. When this material was used to perform the alkylation of 2,6-dimethylphenol with iodopropane under normal stirred conditions the reaction was found to proceed very rapidly initially but the rate tailed off at about 85% completion. By applying ultrasound to the system at this point the reaction could be rapidly forced to completion. Such results support the hypothesis that sonication has a two-fold effect on heterogeneous reactions namely (a) a reduction in particle size affording a far larger surface area for reactivity and (b) much more efficient reagent mixing and mass transfer than can be achieved by normal stirring.

This methodology was found to be highly successful when applied to the *O*-alkylation of 5-hydroxychromones, reactions which are normally very difficult to perform due to electronic hindrance to the OH group caused by hydrogen bonding between it and the carbonyl group [54]. For example the alkylation of 5-hydroxy-2-carboxyethyl-4H-1-benzopyran-4-one to (**29**, R=$CH_2CH_2CH_3$) with 1-iodopropane at 65°C in NMP is only 28% complete after 1.5 h (eqn 3.29) whereas under ultrasonic irradiation complete reaction was obtained in this time period [54].

(3.29)

(29)

3.2.3   Heterogeneous catalytic reactions

Using a bath

Hydrogenation
It is at first a little surprising to find ultrasound of use in hydrogenation reactions especially when it is recalled that one of the main physical effects of ultrasonic irradiation of a liquid saturated with a gas is efficient degassing. We will see below that ultrasonic enhancement to direct hydrogenation using H_2 gas and a catalyst has been achieved using a probe under positive gas pressure. In the case of ultrasonic baths however enhancements to hydrogenation have only been reported in examples where the hydrogen source is not the gas itself. Boudjouk has shown that formic acid and palladium-on-carbon are an effective couple for the hydrogenation of a wide range of alkenes at room temperature in the presence of low intensity ultrasonic fields (50 kHz) [55]. Thus the conjugated ketone can be hydrogenated specifically at the double bond (100% in 1 h) (eqn 3.30). The facility thus exists for clean and rapid hydrogenations which can be carried out without effecting other functional groups.

$$(3.30)$$

A similar technique employs hydrazine as the source of hydrogen again using palladium-on-charcoal as catalyst in ethanol solvent. Results obtained at room temperature for the total hydrogenation of diphenylacetylene (2 h) appear to be almost as good as those obtained under reflux and are substantially better than obtained with stirring at 25°C (6 h) (eqn 3.31) [56].

$$Ph - C \equiv C - Ph \quad \longrightarrow \quad Ph\,CH_2CH_2\,Ph \qquad (3.31)$$

Phase transfer catalysis
The use of ultrasound to produce either extremely fine emulsions (from immiscible liquids), or efficient mass transfer and surface activation (in solid/liquid systems) has led to the use of ultrasound to enhance or even replace phase transfer catalysts (PTC). The normal function of a PTC is to enable the transfer of a water soluble reagent into an organic phase where it can more readily react with an organic substrate. As an example of phase transfer catalysis, consider the quaternary ammonium salt $[(C_8H_{17})_3N^+CH_3]Cl^-$. This salt is itself quite soluble in non-polar (organic) media owing to the presence of the long hydrocarbon chains. When this PTC is dissolved in $CHCl_3$ and the solution shaken with aqueous NaOH some of the OH^- ions in the aqueous phase will be partitioned (pulled) into the organic phase by the PTC in exchange for its own Cl^- ions which are transferred back into the water. Hydroxide ion in chloroform is a powerful base and, under these conditions it can remove a proton from the solvent to give $^-CCl_3$ which, in turn, decomposes to the reactive electrophilic intermediate $:CCl_2$ (dichlorocarbene). If an alkene e.g. styrene, is present in the organic phase the carbene adds to it and the product of the reaction is a dichlorocyclopropane (see eqn 3.21 above). The ability to increase the range of reactivity of a species by transferring it from one phase to another is a powerful tool in chemical synthesis. There are however two drawbacks to the use of phase transfer catalysts (i) the PTC itself is generally expensive and (ii) a PTC is potentially dangerous in that it will, by its very nature, catalyse the transfer of chemicals into human tissue. The potential health hazard of using a PTC must however be counterbalanced by the synthetic efficiency which follows from its employment since without the addition of a PTC many heterogeneous reactions are limited to the interfacial regions of contact and are consequently slow.

It is liquid–liquid reactions involving phase transfer catalysts which generally benefit from the use of ultrasound. Sonication produces homogenisation — i.e. very fine emulsions — which greatly increase the reactive interfacial area and allows faster reaction at lower temperatures. Davidson has reported an example of this with the ultrasonically enhanced saponification of wool waxes by aqueous sodium hydroxide using tetra *n*-heptylammonium bromide as a PTC [57].

In another example increased yields of products are obtained under ultrasonic irradiation in the PTC alkylation of the isoquinoline derivatives (**30**) using 50% aqueous NaOH as base [58]. Efficient mixing is not easy to achieve for this system under normal reaction conditions due to the viscosity of the aqueous base. In the specific case of alkylation with benzyl chloride ultrasound plus $[Et_3NCH_2Ph]^+Cl^-$ achieved 60% yield in 20 minutes compared with only 50% in 2 h with stirring (eqn 3.32).

$$(3.32)$$

Rate enhancements have also been reported in several solid/liquid heterogeneous substitutions involving the use of phase transfer catalysis (PTC). In several instances the use of ultrasound has enabled cheaper PTC reagents to be used. For example the efficiency of *N*-alkylation of amines with alkyl halides in toluene in the presence of solid KOH using the inexpensive and non-toxic PEG methyl ether as PTC is markedly increased (eqn 3.33) with ultrasound. When R = benzyl and $R^1 = R^2$ = phenyl the yield is 98% in 1 h at 25–50°C under sonication compared with only 70% after 48 h under reflux. In the absence of PTC however there was no reaction under sonication. This result emphasises that the increase in reactivity is not simply a matter of increasing interfacial contact area. In this situation the ultrasonic acceleration was ascribed to a combination of fragmentation of solid KOH and efficient mixing, enabling better mass transfer.

$$RX \;+\; R^1R^2NH \longrightarrow R\,R^1R^2N$$

$$(3.33)$$

Attempted autoxidation of 4-nitrotoluene (eqn 3.34) in the presence of O_2 using KOH and PEG 400 as PTC gives a mixture comprising entirely of the dimers (**31**) and (**32**) with combined yields ranging from 19.3 to 38.7% depending on O_2 pressure and reaction conditions [60]. Sonication not only reduces reaction times by a factor of three but, more significantly, effects the selectivity of such reactions. At the same temperature (25°C) and with reaction times of 1 h the product distribution of the reaction was changed by irradiation such that 4-nitrobenzoic acid (**33**) was the main product with yield % acid as high as 43.1% (with only 2.3% total dimer) depending on conditions.

(3.34)

(33)   (31)   (32)

Catalysis using solid supports

Ando has reported the efficient preparation of benzyl cyanide from benzyl bromide (90%) by direct reaction of the halocompound with KCN in toluene in the presence of activated alumina and ultrasound (there is very little reaction in the absence of ultrasound) (eqn 3.35). This is an interesting reaction in that it is representative of a range of supported reactions. The displacement does not occur in the absence of alumina and a small amount of water seems to be critical for high yields. Both of these factors suggest that the effect of ultrasound is concerned with the forcing of KCN on to the alumina surface which itself is activated by the presence of traces of water.

$$PhCH_2Br + KCN \xrightarrow[PhCH_3]{Al_2O_3} PhCH_2CN + KBr \qquad (3.35)$$

It was as a result of further work on this reaction that Ando achieved a real goal in sonochemistry — the use of ultrasound to switch a reaction pathway [61]. His studies of the reaction between 4-methylbenzyl bromide and toluene in the presence of potassium cyanide and alumina (Scheme 3.10) revealed that under ultrasonic conditions the reaction produces the benzyl cyanide (**34**, 77%) in 24 h, as expected from the above. In contrast, after the same time, but without sonication, the main product was the Friedel–Crafts adduct (**35**, 75%). The explanation advanced is that ultrasound increases contact between cyanide ion and alumina which decreases the Friedel–Crafts activity of the alumina surface while promoting the nucleophilicity of the adsorbed cyanide.

A classical method for the preparation of amino acids is the Strecker synthesis. Recently a new approach to this synthesis has been reported in which amino nitriles have been prepared in high yield by the direct reaction between an aldehyde, KCN and NH$_4$Cl in acetonitrile (eqn 3.36) [62]. In the case of benzaldehyde the yield of the target amino nitrile is poor under normal stirred conditions with starting material, benzoin and hydroxynitrile predominating (Table 3.5). The presence of alumina suspended in acetonitrile increases the proportion of amino nitrile but the results make it clear that the optimum reaction conditions require that in addition to

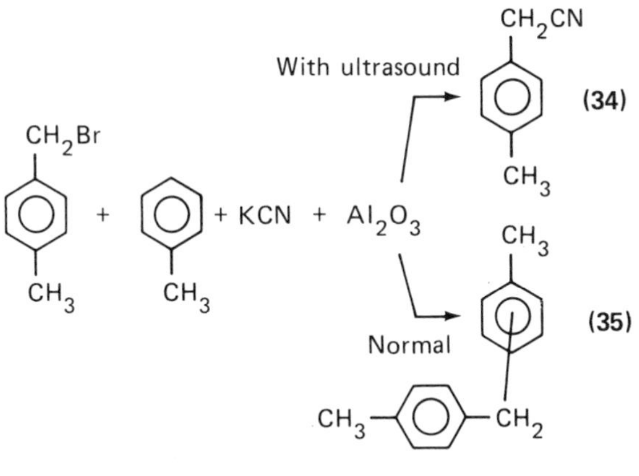

Scheme 3.10

Table 3.5 — Strecker synthesis of α-aminonitriles

	Reaction products in CH_3CN at 50°C			
	PhCHO	Ph—CH—CN \ OH	Ph—CH—CN \ NH_2	Ph—CH—CO—Ph \ NH_2
Stirred	35	38	6	21
Stirred+Al_2O_3	8	19	64	9
Sonicated	11	45	23	22
Sonicated+Al_2O_3	0	3	90	7

suspended alumina, sonication is required. Under these conditions the yield of the amino nitrile reaches 90%.

$$Ph\,CHO \;+\; KCN \;+\; NH_4Cl \;\longrightarrow\; \underset{NH_2}{Ph\text{-}CH\text{-}CN} \qquad (3.36)$$

Basic alumina is an effective catalyst for some condensation reactions and has been used for the synthesis of 3-nitro-2H-chromones (eqn 3.37). The reaction was found to be effective in the absence of solvent but even so the reaction times were relatively long. The reaction times could however be reduced to 2–4 hours when a solid mixture of alumina, onto which the reagents had been adsorbed, was subjected to ultrasonic irradiation in a cleaning bath at 30–35°C [63]. By this method (36) could be prepared (85% based on nitrostyrene) in 2 h.

(3.37)

(36)

Using a probe

Hydrogenation

Suslick has reported the sonochemical activation of nickel powders as hydrogenation catalysts. Normally simple nickel powder is a reluctant catalyst for the hydrogenation of alkenes yet ultrasonic irradiation offered a reactivity comparable with Raney nickel [64]. In this case, ultrasound gave an unexpected decrease in surface area due to aggregation of particles, with electron micrographs indicating a smoothing of the nickel suface. Auger electron spectroscopy revealed an increase in the nickel/oxygen ratio at the surface. The explanation suggested was that abrasion from interparticle collisions removes the oxide layer of the nickel giving the observed increased reactivity.

A commercially exploited example of a sonochemically enhanced catalytic reaction is the ultrasonic hydrogenation of soyabean oil [65]. A three phase non-aqueous system is used comprising liquid oil, H_2 gas and solid catalyst. The catalyst was either 1% copper chromite or 0.1% Nysel (25% nickel) at 115 psig and 180°C. The method employs a flow through cell operating at 0.5–2.0 litres/hour providing sonication at 20 kHz and leading to a tube reactor of dimension 120 feet x 1/8 inch. This has considerable advantages over the currently used batch methods which require much longer reaction times.

Alkene isomerisation

The sonolysis of metal carbonyls is a study to which reference has already been made (Scheme 3.3), but within this area lies the potentially important topic of sonochemically produced homogeneous catalysis [18]. The transient coordinatively unsaturated species generated during the sonolysis of $Fe(CO)_5$, $Fe_2(CO)_9$ and $Fe_3(CO)_{12}$ in the presence of pent-1-ene each gave trans- and cis-pent-2-ene in the thermodynamic ratio of 3.5:1, a result indicating that the same catalytic species was present in each case. In contrast to this result the sonolysis of $Ru(CO)_5$ in pent-1-ene produced a different trans:cis ratio of 2.8:1 giving an indication of the potential of this new method of catalyst activation.

REFERENCES

[1] E. C. Couppis and G. E. Klinzing, *AIChE J.*, 1974, **20**, 485.
[2] S. Folger and D. Barnes, *Ind. Eng. Chem. Fundam.*, 1968, **7**, 222.
[3] Special edition of the journal 'Ultrasonics' Covering the R.S.C. Sonochemistry Symposium, Warwick 1986, *Ultrasonics*, 1987, **25**, January.

[4] J. Lindley, P. J. Lorimer and T. J. Mason, *Ultrasonics*, 1986, **24**, 292.

[5] J. W. Chen and W. M. Kalback, *Ind. Eng. Chem. Fundam.*, 1967, **6**, 175.

[6] D. S. Kristol, H. Klotz and R. C. Parker, *Tetrahedron Letts.*, 1981, **22**, 907.

[7] T. J. Mason and J. P. Lorimer, *J. Chem. Soc., Chem. Commun.*, 1980, 1135.

[8] T. J. Mason, J. P. Lorimer and B. P. Mistry, *Tetrahedron Lett.*, 1982, **23**, 5563.

[9] T. J. Mason, J. P. Lorimer and B. P. Mistry, *Tetrahedron*, 1985, **26**, 5201.

[10] T. J. Mason, J. P. Lorimer and B. P. Mistry, *Ultrasonics International 85, Proceedings*, Butterworth, UK, 1985, 839.

[11] C. J. Burton, *J. Acoust. Soc. Am.*, 1948, **20**, 186.

[12] K. Makino, M. M. Mossoba and P. Riesz, *J. Phys. Chem.*, 1983, **87**, 1369.

[13] R. Schultz and A. Henglein, *Z. Naturforsch.*, 1953, **8**, 160.

[14] A. Henglein and C. H. Fischer, *Ber. Bunsenges Phys. Chem.*, 1984, **88**, 1196.

[15] K. S. Suslick, J. J. Gawlenowski, P. F. Schubert and H. H. Wang, *J. Phys. Chem.*, 1983, **87**, 2299.

[16] K. S. Suslick, D. A. Hammerton and R. F. Cline Jr., *J. Am. Chem. Soc.*, 1986, **108**, 5641.

[17] A. M. Horton, D. M. Hollinshead and S. V. Ley, *Tetrahedron*, 1984, **40**, 1737.

[18] K. S. Suslick, J. W. Goodale, P. F. Schubert and H. H. Wang, *J. Am. Chem. Soc.*, 1983, **105**, 5781.

[19] K. S. Suslick and P. F. Schubert, *ibid.*, 1983, **105**, 6042.

[20] J.-L. Luche and J. C. Damanio, *J. Am. Chem. Soc.*, 1980, **103**, 7926.

[21] J. D. Sprich and G. S. Lewandos, *Inorg. Chim. Acta.*, 1982, **76**, 1241.

[22] T. Azuma, S. Yanagida, H. Sakurai, S. Sasa and K. Yoshino, *Synth. Commun.*, 1982, **12**, 137.

[23] R. D. Rieke, *Acc. Chem. Res.*, 19??7??, ,**10** 301.

[24] P. Boudjouk, D. P. Thompson, W. H. Ohrbom and B. H. Han, *Organometallics*, 1986, **5**, 1257.

[25] H. Bonnermann, B. Bogdanovic, R. Brinkman, D. W. He and B. Spliethoff, *Angew. Chem., Int. Ed. Engl.*, 1983, **22**, 728.

[26] W. Oppolzer and P. Schneider, *Tetrahedron Lett.*, 1984, **25**, 3305.

[27] J.-L. Luche, C. Petrier and C. Dupuy, *Tetrahedron Lett.*, 1984, **25**, 753.

[28] T. S. Chou and M. L. You, *ibid.*, 1985, **26**, 4495.

[29] C. Petrier, A. L. Gemal and J.-L. Luche, *ibid.*, 1982, **23**, 3361.

[30] B. H. Han and P. Boudjouk, *J. Org. Chem.*, 1982, **47**, 5030.

[31] B. H. Han and P. Boudjouk, *J. Org. Chem.*, 1982, **47**, 751.

[32] N. Joshi and H. M. R. Hoffmann, *Tetrahedron Lett.*, 1986, **27**, 687.

[33] G. Mehta and H. S. P. Rao, *Synth. Comm.*, 1985, **15**, 991.

[34] J. Yamashita, Y. Inoue, T. Kondo and H. Hashimoto, *Bull. Chem. Soc. Jpn.*, 1984, **57**, 2335.

[35] O. Repic and S. Vogt, *Tetrahedron Lett.*, 1982, **23**, 2729.

[36] O. Repic, P. G. Lee and N. Giger, *Org. Prep. Proc. Int.*, 1984, **16**, 25.

[37] T. Kitazume and N. Ishikawa, *J. Am. Chem. Soc.*, 1985, **107**, 5186.

[38] A. Solladie-Cavallo, D. Farkharic, S. Fritz, T. Lazrak and J. Suffert, *Tetrahedron Lett.*, 1984, **25**, 4117.

[39] J.-L. Luche, C. Petrier, J. P. Lansard and A. E. Greene, *J. Org. Chem.*, 1983, **48**, 3837.

[40] C. Petrier, J.-L. Luche and C. Dupuy, *Tetrahedron Lett.*, 1984, **25**, 3463.

[41] J. Lindley, P. J. Lorimer and T. J. Mason, *Ultrasonics*, 1986, **24**, 292.

[42] K. S. Suslick and R. E. Johnson, *J. Am. Chem. Soc.*, 1984, **106**, 6856.

[43] A. Fuentes and J. V. Sinisterra, *Tetrahedron Lett.*, 1986, **27**, 2967.

[44] H. C. Brown and U. S. Racherla, *Tetrahedron Lett.*, 1985, **26**, 2187.

[45] S. L. Regen and A. Singh, *J. Org. Chem.*, 1982, **47**, 1587.

[46] T. Ando, T. Kawate, J. Yamawaki and T. Hanafusa, *Synthesis*, 1983, 637.

[47] H. Priebe, *Acta Chem. Scand., Ser B*, 1984, **38**, 895.

[48] J. Mertens, W. Vanryckeghem, A. Bossuyt, P. Vden Winkel and R. Vandendriessche, *J. Labelled. Cpd. Radiopharm.*, 1984, **21**, 843.

[49] B. H. Han and P. Boudjouk, *Tetrahedron Lett.*, 1982, **23**, 1643.

[50] E. Lukevics, V. N. Gevorgyan and Y. S. Goldberg, *ibid.*, 1984, **25**, 1415.

[51] J. Yamakawi, S. Sumi, T. Ando and T. Hanafusa, *Chem. Lett.*, 1983, 379.

[52] S. Moon, L. Duchin and J. Cooney, *Tetrahedron Lett.*, 1979, 3917.

[53] T. J. Mason, J. P. Lorimer, A. Moore, A. T. Turner and A. R. Harris, *Ultrasonics Int. 87, Conf. Proc.*, 1988, 767.

[54] T. J. Mason, J. P. Lorimer, A. T. Turner and A. R. Harris, *J. Chem. Res. (S)*, 1988, 80.

[55] P. Boudjouk and B. H. Han, *J. Catal.*, 1983, **79**, 489.

[56] D. H. Shin and B. H. Han, *Bull. Korean Chem. Soc.*, 1985, **6**, 247.

[57] R. S. Davidson, A. Safdar, J. D. Spencer and D. W. Lewis, *Ultrasonics*, 1987, **25**, 35.

[58] J. Ezquerra and J. Alvarez-Builla, *J. Chem. Soc., Chem. Commun.*, 1984, 54.

[59] R. S. Davidson, A. M. Patel, A. Safdar and D. Thornthwaite, *Tetrahedron Lett.*, 1983, **24**, 5907.

[60] R. Neumann and Y. Sasson, *J. Chem. Soc., Chem. Commun.*, 1985, 616.

[61] T. Ando, S. Sumi, T. Kawate, J. Ichihara and T. Hanafusa, *J. Chem. Soc., Chem. Commun.*, 1984, 439.

[62] T. Hanafusa, J. Ichihara and T. Ashida, *Chem. Letts.*, 1987, 687.

[63] R. S. Varma and G. W. Kabalka, *Heterocycles*, 1985, **23**, 139.

[64] K. S. Suslick and D. J. Casadonte, *J. Amer. Chem. Soc.*, 1987, **109**, 3459.

[65] K. J. Moulton, S. Koritala and E. N. Frankel, *J. Am. Oil Chem. Soc.*, 1983, **60**, 1257.

4

Polymers

4.1 INTRODUCTION

In this chapter we will discuss the effect of low frequency (<400 kHz) high intensity (>3 W cm^{-2}) ultrasonic waves on macromolecules and examine how the parameters such as frequency, intensity, hydrostatic pressure etc. affect both the depolymerisation and polymerisation processes. High frequency (>1 MHz) low intensity waves which are useful in providing information on relaxation phenomena such as segmental motion, conformational change, vibrational-translational energy interchange and polymer–solvent interaction will be dealt with in Chapter 6.

4.2 DEGRADATION OF POLYMERS

It is now well established that the prolonged exposure of solutions of macromolecules to high energy ultrasonic waves produces permanent reductions in the solution viscosity (Table 4.1).

Table 4.1 — Flow times of 1% solution of polystyrene in toluene (air saturated) versus insonation time (960 kHz, 6.8 W cm^{-2})

Irradiation time (minutes)	0	30	60	90	120
Flow time (seconds)	23.6	19.7	19.0	18.4	18.3

Even when the irradiated polymers are isolated and redissolved their viscosities remain low in comparison to those of the non-irradiated solutions. If these observations are the result of cavitation then the extent of degradation for any system should be zero if the process is carried out under conditions which eliminate cavitation — i.e. in either vacuo or at high overpressure. These conditions are achieved by degassing the solution, to remove gas nuclei which subsequently could have led to a cavitational event, or increasing the external pressure — see 2.4.4. In contrast to this expectation early workers such as Schmid [1] and Mark [2] observed degradation both *in vacuo* and at a high overpressure. For example, Schmid studying the degradation of polystyrene in benzene at an acoustic pressure equivalent to 5 atm still observed degradation under 15 atm of applied pressure. Such an overpressure, in Schmid's view, ought to have been sufficient to inhibit the degradation. To explain these observations, Schmid and Mark suggested that the decreases in solution viscosity on irradiation were the consequence of degradation of the polymer, as a result of increased frictional forces between the ultrasonically accelerated faster moving solvent molecules and the larger less mobile macromolecules. Although envisaging different modes of interaction between the solvent and polymer molecules both investigators concluded that the increased forces were sufficient to break a C–C bond.

Schmid for his part considered two ideal cases for the degradation of the polymer in solution; the first in which the macromolecules were considered to be immobile in solution and the solvent molecules were swept past them under the action of the applied acoustic field; the second involved an allowance for macromolecule movement, the macromolecule moving with the solvent under the action of the applied ultrasonic field.

In the case of an immobile macromolecule in solution it is possible to estimate a value of the frictional force developed between the solvent and polymer molecules by assuming the macromolecule to consist of a number (n) of solid spherical entities and applying the Stokes' formula in a modified form (eqn 4.1).

$$F = 6\eta r V n \qquad (4.1)$$

where η is the fluid viscosity, r is the radius of the sphere and V is the velocity. For example, for a polystyrene molecule (RMM $\sim$ 300 000; $n \sim$ 3000) in benzene as solvent (viscosity $\sim 6.2 \times 10^{-4} \mathrm{N\,s\,m^{-2}}$), where the pendant benzene rings can be assumed to be acting as the spheres ($r = 3 \times 10^{-10} \mathrm{m}$), the total frictional force developed for a fluid flowing under the action of ultrasonic field with a velocity of 0.5 ms^{-1} can be estimated to be 5.3×10^{-9} newtons [i.e. $F = (6\pi) \times (6.2 \times 10^{-4}) \times (3 \times 10^{-10}) \times (50 \times 10^{-2}) \times 3000$].

From spectroscopic data it has been deduced that the value needed to effect breakage of a C–C bond is 4.5×10^{-9} N. Since the value estimated from the frictional approach is some 20% higher it could be inferred that Schmid's frictional approach to polymer degradation is in fact correct. If so, then there ought not to be any significant degradation in a system where the polymer and solvent molecules have similar densities. This is not the case as Table 4.2 shows, although there is a slight density dependence.

Further evidence that this model cannot be used to predict absolutely the course of degradation is afforded by the influence of changes in irradiation frequency on degradation. If, as Schmid suggested, degradation is the result of solvent movement past the macromolecule, then for a rigid macromolecular structure a decrease in frequency should allow the macromolecule more time to accommodate the impact of the solvent — i.e. degradation ought to decrease with decrease in frequency. In

Table 4.2 — Degradation of polystyrene in CCl$_4$/toluene mixtures at 55°C

Time (minutes)	Solvent density (g cm^{-3})				
	0.867	1.058	1.234	1.454	1.601
0	208 000	203 000	197 000	190 000	189 000
10	140 000	158 000	157 000	151 000	133 000
30	100 000	124 000	129 000	112 000	109 000
60	80 000	100 000	108 000	89 000	93 000
120	61 000	78 000	85 000	73 000	80 000

practice the opposite effect is observed — degradation increases with decrease in frequency (Fig. 4.1).

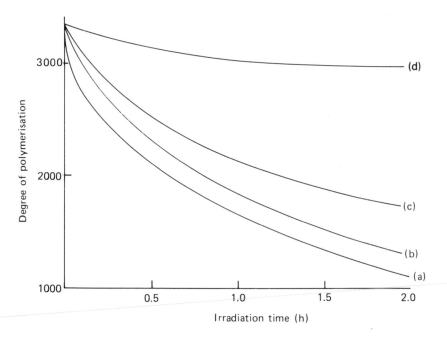

Fig. 4.1 — Effect of frequency on the degradation of polystyrene (1%) in benzene. (a) 1 MHz; (b) 1.25 MHz; (c) 1.5 MHz; (d) 2 MHz.

The second ideal case identified by Schmid assumed that the macromolecule moved with the solvent and in this case equation (4.2) must be employed:

$$F = \frac{nmwV}{N_A} \tag{4.2}$$

where m is the RMM of the monomer, N_A is the Avogadro number and w is the angular frequency of the applied ultrasonic wave. For an applied frequency of 500 kHz, the frictional force developed in the above polymer system, according to equation (4.2), may be calculated to be 8.1×10^{-16} N. This is obviously many orders of magnitude less than that required to rupture a C–C bond. The conclusion must be that the more dilute a polymer solution, the more likely it is that the macromolecule will move with the solvent since there is less interaction, and the smaller will be the depolymerisation effect. This is certainly not observed in practice (see 4.3.7) and identifies a flaw in Schmid's argument. Further this second model, like the first implies (eqn 4.2) that polymer degradation increases with increase in frequency. As already pointed out (Fig. 4.1) this is not observed experimentally — see 4.3.1 below.

Mark's hypothesis, although still suggesting that frictional forces were the origin of the degradation, took into account the molecular configuration of the polymer. It is well known that macromolecules in solution are neither completely stretched out in a straight line (Fig. 4.2a), nor are they completely coiled up in a ball (Fig. 4.2b), but exist in an intermediate shape of moderately undulated configurations (Fig. 4.2c).

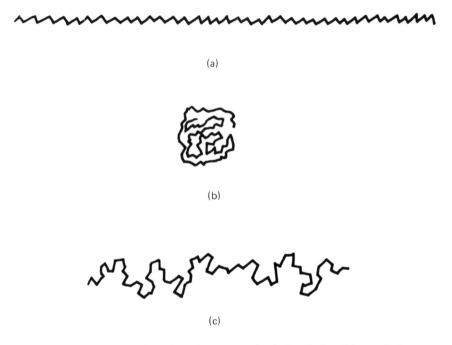

(a)

(b)

(c)

Fig. 4.2 — The possible configuration of a macromolecule in solution: (a) stretched out; (b) coiled up; (c) intermediate.

Under the action of an applied acoustic field the suggestion was that there would be regions within the polymer where rotation (and vibration) of individual segments were able to take place freely, in phase with the rapid oscillatory movement of the

solvent. This segmental movement (termed micro Brownian motion) was in addition
to the movement of the macromolecule as a whole (macro Brownian motion).
However, in that segmental motion is a cooperative effect and depends upon the
interaction between various units in the macromolecule, there will be regions where
movement lags well behind the motion of the solvent molecules leading to a
concentration of stress and consequent rupture of a C–C bond. Estimates by Mark of
the frictional forces between the macromolecules and the solvent molecules
provided values which were large enough to break not only C–C bonds, but also
C=C and C=O bonds.

Despite the inability to account for several of the experimental observations, the
above theories predict that the frictional forces developed are dependent upon the
size of the macromolecule and that the rate of degradation decreases with decrease in
chain size (Figs 4.3 and 4.4).

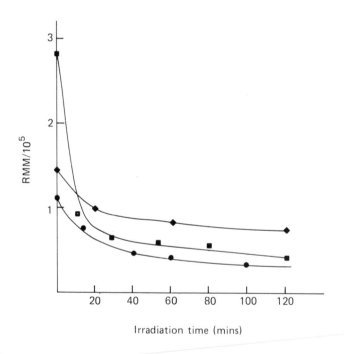

Fig. 4.3 — Degradation rate versus initial RMM (polystyrene in toluene, 70°C).

Based upon experimental data, Schmid was able to show this mathematically for
dilute solutions (<0.02 M) as

$$dx/dt = k(P_t - P_L) \tag{4.3}$$

and for concentrated solutions as

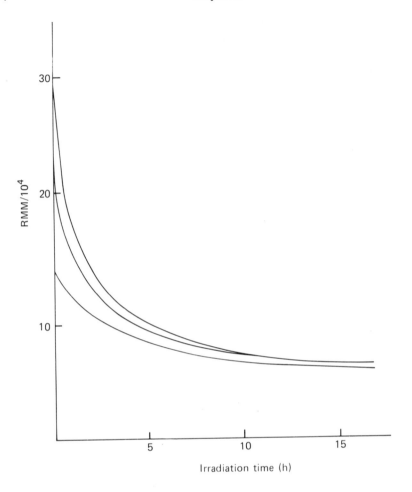

Fig. 4.4 — Depolymerisation of polystyrene in benzene ($f = 750$ kHz; Intensity = 12 W cm^{-2}).

$$\mathrm{d}x/\mathrm{d}t = k \log (P_t/P_o) \tag{4.4}$$

where dx is the number of chemical bonds broken in unit volume in an irradiation time dt, P_o is the initial degree of polymerisation, P_t is the degree of polymerisation at time t and P_L is the limiting degree of polymerisation — i.e. the lowest value to which the polymer could be degraded.

The failure of the above theories to predict, qualitatively at least, the course of degradation under certain conditions is most certainly based upon the assumption that cavitation did not play a significant role in ultrasonic degradation.

There is now a wealth of experimental information [1,3–11] to suggest that degradation is due to cavitation effects. For example:

(a) The ultrasonic degradation of polystyrene in benzene increases with a decrease in the applied ultrasonic frequency (see 2.6.1).
(b) The ultrasonic degradation of aqueous poly(acrylic acid) decreases with the addition of ether, i.e. increased vapour pressure (see 2.6.2).
(c) The ultrasonic degradation of polystyrene in toluene decreases with an increase in the reaction temperature (see 2.6.3).
(d) The ultrasonic degradation of an air-saturated toluene solution of polystyrene is greater than when the solution is degassed.
(e) The ultrasonic degradation of polystyrene in toluene increases with an increase in the externally applied pressure.

What is debatable however, is whether the degradation is caused by

(i)   The hydrodynamic forces of cavitation — i.e. the shock wave energies produced on bubble implosion.
(ii) The shear stresses at the interface of the pulsating bubbles.
(iii) The associated thermal and pressure increases within the bubbles themselves.

All the above are dependent upon the same factors — i.e. intensity, frequency, gas content and type etc. The current view is that ultrasonic degradation is for the main part mechanical in its origin and due to the high pressures associated with the collapse of the bubble. There may be the possibility that part of the degradation is thermal but this would only be significant for those macromolecules with a sufficiently high vapour pressure to allow entry into the cavitation bubble.

4.3   FACTORS AFFECTING POLYMER DEGRADATION

4.3.1   Frequency

One of the earliest reports on the effect of frequency was by Schmid. Working at a constant ultrasonic intensity of 1 W cm^{-2} and irradiation frequencies of 300, 175 and 10 kHz, Schmid concluded that the frequency had practically no effect on the degradation of poly(methyl methacrylate) in benzene (0.3% w/v) — Fig. 4.5.

However a closer examination of Fig. 4.5 seems to indicate that the degradation rate, and to a lesser degree the extent of degradation, decreases with an increase in the irradiation frequency. The general conclusion that the higher the irradiation frequency the lower is the degradation rate, and the higher the limiting degree of polymerisation, P_L, is a view now shared by most workers in the field. Perhaps the only contentious point is the extent to which a polymer is degraded at a particular frequency. We have already seen (2.3) that the higher the frequency of the ultrasonic wave the more rapidly it is attenuated. It is important, therefore, to recognise that when comparing the effects of frequency, intensity must be held constant. For example, Gaertner observed depolymerisation at both 400 kHz and 2500 kHz, the lower frequency necessitating an intensity of only 0.5 W cm^{-2} whereas the higher frequency required approximately 2 kW cm^{-2}. Unfortunately many workers in the field do not report the acoustic power entering the system but only the power delivered to the transducer. However what is certain is that the higher the applied frequency the shorter is the period in which bubble growth and collapse can occur,

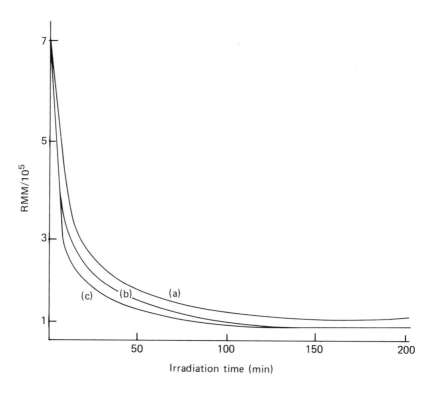

Fig. 4.5 — Degradation of polymethyl methacrylate in benzene as a function of frequency (Intensity = 1 W cm^{-2}; concentration = 0.3% w/v.) (a) f = 300 kHz; (b) f = 10 kHz; (c) f = 175 kHz.

and the more likely it is that there is insufficient time to produce cavitation. According to Neppiras and Noltingk there is a frequency limit (~ 10 MHz) above which cavitation does not occur. This limit depends upon the initial size of the bubble, the external (hydrostatic) pressure (P_h) and the pressure amplitude (P_A) of the ultrasonic wave. Some workers have suggested that there is an optimum frequency at which the cavitation intensity attains a maximum value and at this frequency maximum degradation occurs. For example Mostafa has observed (Fig. 4.1) maximum degradation at 1 MHz for polystyrene in benzene (I = 12.5 W cm^{-2}). In Mostafa's view, the small observed degradation at the highest frequency (2 MHz), where presumably cavitation is negligible, is probably due to resonance effects of some kind. Whatever the form of resonance, it certainly cannot be between the frequency of the molecular vibrations and the frequency of the applied ultrasound since the former (estimated to be 10^8 MHz) is many orders of magnitude greater than the latter (max. 1 MHz). One possible explanation is that degradation is caused by the shear forces set up by the ultrasonic waves. As the frequency, and hence attenuation, of the wave decreases, the shear forces will be complemented by the increased cavitational effects.

4.3.2 Solvent

Most of the experimental data indicate that there is a decrease in the extent of depolymerisation with increasing solvent vapour pressure. For example, Schmid and Rommel observed slower degradation rates in benzene than in toluene for samples of polystyrene and various polyacrylates. Although at 20°C the viscosities of benzene and toluene are similar, 0.65 and 0.59 cP respectively, their vapour pressures differ appreciably being 68.2 mmHg and 25.1 mmHg respectively. Similar trends have been observed for the degradation of nitrocellulose in several solvents, the extent of degradation increasing in the order acetone, ethyl acetate, propyl acetate, isobutyl acetate, isoamylacetate and ethyl benzoate. Although the amount of degradation increases with decrease in the solvent vapour pressure, as predicted (see General Principles, Section 2.6.2), the order is also that of increasing solvent viscosity (Table 4.3).

Table 4.3 — Values of vapour pressure, viscosity and surface tension of various solvents at 20°C

Solvent	Vapour pressure (mmHg)	Viscosity (cP)	Surface tension $(Nm^{-1}/10^{-3})$
Acetone	188.0	0.326	23.7
Ethyl acetate	77.1	0.455	23.9
n-Propyl acetate	28.4	0.590	24.3
i-Butyl acetate	17.6	0.732	—
i-Amyl acetate	5.0	0.872	—
Ethyl benzoate	0.3	2.24	35.5
Benzene	68.2	0.65	28.9
Toluene	25.1	0.59	28.5
CCl$_4$	92.7	0.97	27.0

Thus while vapour pressure may be the major solvent factor involved in the degradation process there could also be a contribution from solvent viscosity or even, yet less likely, from surface tension. It has already been argued (see 2.6.2) that although an increase in viscosity raises the cavitation threshold, (i.e. makes cavitation more difficult), provided cavitation occurs, the pressure effects resulting from bubble collapse will be greater. In other words the cavitational effects of increasing vapour pressure and viscosity are partially compensatory — i.e. act in opposite directions.

An example of this effect may be seen by considering the data of Schmid and Beuttenmuller for the degradation of polystyrene in CCl$_4$/toluene mixtures of various composition (Table 4.4). On proceeding across the composition range from toluene to a 50/50 (v/v) mixture of CCl$_4$ and toluene, the increase in vapour pressure is accompanied by an almost proportional decrease in the degradation rate, measured as the percentage decrease in RMM per minute.

Table 4.4 — Degradation of polystyrene in various CCl_4/toluene mixtures

% CCl_4	0	25	50	80	100
Degradation rate (% min^{-1})	5.5	3.1	2.7	2.6	4.8
†Vapour pressure at 55°C (mmHg)	115	189	254	330	379
Viscosity at 55°C (cP)	0.4	—	—	—	0.6

†Calculated from $P = x_A P_A + x_B P_B$.

It is surprising that further additions of CCl_4 (80% and 100%), which increase the vapour pressure, do not lead to a decrease in the degradation rate. On the contrary, the degradation for pure CCl_4 is almost identical to that of pure toluene even though the vapour pressure has increased more than threefold. The viscosity, however, has increased by approximately 50% with change of solvent.

In contrast to the idea that vapour pressure is the major solvent parameter in determining the extent of degradation, Doulah has suggested it depends upon the solvent's solvating capacity. His proposal is that the energy released by the imploding cavitation bubbles, in the form of hydrodynamic shock waves, is more effective in 'good' solvents, where the chains are relatively extended (Fig 4.2c), than in 'poor' solvents (Fig. 4.2b).

4.3.3   Temperature
An alternative method of raising the vapour pressure of a solvent is to increase the experimental temperature. The consequence should be both a decrease in the rate of degradation and an increase in the limiting degree of polymerisation (i.e. higher final RMM value) as a result of the lower intensities of cavitational collapse at the higher temperatures (see 2.6.2). Table 4.5 and Fig. 4.6 show these predictions are borne out in practice.

Table 4.5 — Percentage ultrasonic degradation and final RMM of polystyrene in toluene at various temperatures

Temperature	40	60	80	100	120
% degradation/(h)	79	78	71	57	32
RMM (10^3)	80	65	128	270	—

4.3.4   Nature of the gas
Whereas all workers agree that the extent of degradation is increased in the presence of gas, there is some dispute regarding the extent to which it takes place *in vacuo*. For example Weissler (Fig. 4.7) and separately Prudhomme and Grabar (Table 4.6),

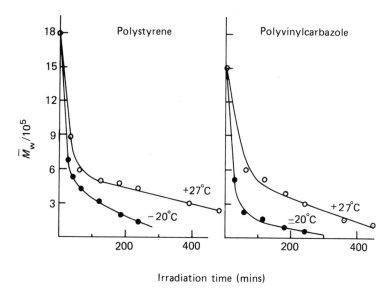

Fig. 4.6 — Effect of temperature on degradation rate.

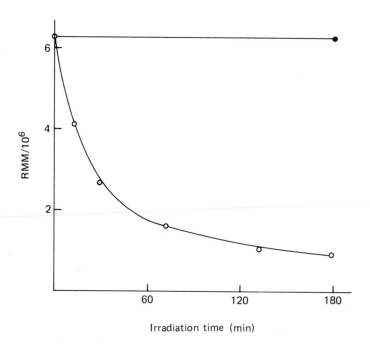

Fig. 4.7 — Ultrasonic degradation of an air-saturated (○) and degassed (●) polystyrene sample in toluene (0.5%) at 400 kHz.

Table 4.6 — Degradation of polystyrene in toluene (1%) in the presence and absence of a gas

Irradiation time (min)	0	30	60	90	120
Flow time (air saturated) (s)	23.6	19.7	19.0	18.4	18.3
Flow time (degassed) (s)	23.6	24.0	—	—	23.6

investigating the degradation of polystyrene in toluene, failed to observe any appreciable degradation in the absence of gas.

Melville and Murray on the other hand, observed little if any degradation for poly(methyl methacrylate) samples of moderate RMM, but noted that a high RMM sample was readily broken down (Fig. 4.8).

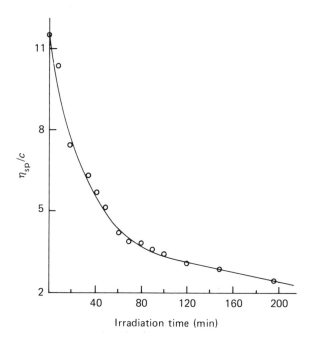

Fig. 4.8 — Degradation of a high relative molar mass sample of poly(methyl methacrylate) in benzene *in vacuo*.

Although at first glance the extent of degradation (Fig. 4.8) appears quite appreciable, a comparison of the intrinsic viscosity after ten minutes *in vacuo* ($[\eta] = 11$) with that for the same solution when irradiated in air ($[\eta] = 0.34$) indicates that the extent of degradation is considerably larger in the presence of gas.

If, according to the ideas presented earlier (2.6.4), cavitation is the origin of degradation, then the extent of degradation should be greatest for gases with the highest specific heat ratio values (γ). This can in part be confirmed by considering Table 4.7 where the intrinsic viscosities (i.e. $[\eta] \propto$ RMM) obtained at several

Table 4.7—Intrinsic viscosity versus irradiation time of polystyrene in benzene (1%)

Dissolved	\multicolumn{7}{c}{Irradiation time (minutes)}						
	0	10	20	40	60	120	240
Air	1.81	1.22	0.95	0.81	0.68	0.60	0.47
N_2	1.82	—	1.03	—	0.8	—	—
O_2	1.79	—	1.04	0.82	0.8	—	—
H_2	1.81	—	1.09	—	0.68	—	—
A	1.81	1.30	1.20	0.95	0.77	0.71	0.56
NH_3	1.81	1.67	1.58	1.28	1.06	—	—
CO_2	1.80	1.66	1.65	1.50	—	—	—
SO_2	1.90	—	1.72	—	1.86	—	—

irradiation time intervals are given for solutions of polystyrene in benzene (1%) which have been saturated with a variety of different gases.

Although for the diatomic gases (N_2, O_2, H_2, air) the degradation rates (Table

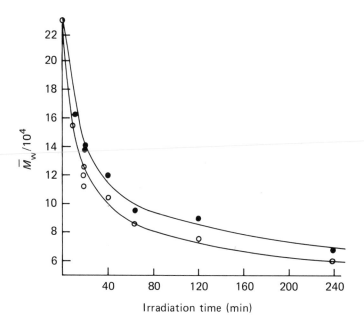

Fig. 4.9 — Degradation of polystyrene in benzene in the presence of argon (●) and air (○).

4.8) as measured by $([\eta]_{t=0} - [\eta]_{t=30})/[\eta]_{t=0}$ are higher than for the polyatomic gases (NH_3, CO_2, SO_2), the rate for argon, a monatomic gas, is surprisingly, lower than for the diatomics (Fig. 4.9).

Jellinek has explained this anomaly in terms of the velocity at which the gas-filled cavities collapse. The larger the velocity of the cavity collapse, the faster the solvent molecules are swept past the polymer molecule and the faster is the degradation rate. Jellinek's estimate of the average collapse velocity of a cavity when filled with a monatomic gas was approximately 70% that of a cavity filled with a diatomic gas — i.e. a slower collapse velocity. It may be fortuitous, but the ratio of the degradation rates for argon and oxygen, gases of similar solubility is 0.8 — i.e. 80%, a value close to the 70% obtained above (see Table 4.8).

According to Jellinek's deductions the differences in degradation rate expected for the diatomic and polyatomic gases should be smaller. A consideration of Table 4.8 shows that this is not observed in practice, the polyatomics having considerably lower rates. In fact, unlike the diatomics, the degradation rates for polyatomics are quite dissimilar, the rates decreasing with increase in solubility. If however solubility was the only other contributing factor to degradation a solution saturated with hydrogen ought to degrade faster and to a greater extent than one saturated with any of the other diatomic gases. This is not the case as Table 4.8 shows.

It may be however that the thermal conductivity of the gas plays some role. The value for hydrogen is somewhat higher than those for the other diatomics indicating that more heat (formed in the bubble during collapse) will be dissipated in the surrounding liquid effectively decreasing the maximum temperature, T_{max}, attainable in the bubble. That is not to say that degradation in the presence of cavitation is thermal in origin as work by Melville has shown. Melville carried out both ultrasonic and thermal degradation of two samples of copolymer of methyl methacrylate and acrylonitrile (molar ratio of methacrylate to acrylonitrile 411:1 and 40:1), and observed that whereas the latter copolymer had the faster thermal degradation rate, in the presence of ultrasound both copolymers showed practically the same rate of degradation. Further, a sample of poly(methyl methacrylate) had the same ultrasonic degradation rate as both of the copolymers — Fig. 4.10.

4.3.5 Intensity

In Chapter 2 we explained why there existed a cavitation threshold, i.e. a limit of sound intensity below which cavitation could not be produced in a liquid. We suggested that only when the applied acoustic amplitude (P_A) of the ultrasonic wave was sufficiently large to overcome the cohesive forces within the liquid could the liquid be torn apart and produce cavitation bubbles. If degradation is due to cavitation then it is expected that degradation will only occur when the cavitation threshold is exceeded. This is confirmed by Weissler who investigated the degradation of hydroxycellulose and observed that the start of degradation coincided with the onset of cavitation (Fig 4.11).

It has also been suggested (Chapter 3) that there is an optimum power which can be applied to a system to obtain the most beneficial effect. Qualitatively it may be suggested that the number of bubbles produced at very high intensities serve to reduce power dissipation by reflecting the sound wave and creating a non-linear response to the increase in intensity. Quantitatively we can argue that at high

Table 4.8 — Degradation rate versus gas parameter

Gas	Air	N_2	O_2	H_2	A	NH_3	CO_2	SO_2
Rate (10^{-2} min^{-1})	47.5	43.0	42.5	40.0	33.7	12.7	8.3	5.0
γ	1.40	1.40	1.40	1.41	1.67	1.31	1.30	1.29
Solubility	0.14	0.11	0.22	0.07	0.24	9.95	2.4	88.0
Conductivity (10^2 W m^{-1} K^{-1})	2.23	2.28	2.33	15.9	1.58	2.00	1.37	0.77

intensities (i.e. large P_A values) the cavitation bubbles have grown so large on rarefaction that there is insufficient time available for collapse during the compression cycle (see 2.6.6). Mark has observed this phenomenon for a sample of polystyrene in toluene. Using a sample of initial RMM. 300 000 he observed that the relative viscosity of the solution after 90 minutes irradiation showed very little reduction beyond an input power of 10 W cm^{-2} — Fig. 4.12.

Others workers who have investigated the effect of intensity on degradation are Melville, Wada and Kanane [12] and more recently Chen Keqiang et al. [13]. Melville investigated the effect of intensity, as measured by the anode current supplied to the oscillator, and observed that the degradation rate and extent of degradation for polystyrene in benzene increased with increase in intensity (Fig. 4.13).

Similar behaviour has been observed in other polymer systems, e.g. Wada and Kanane for the degradation of poly(methyl methacrylate) in chloroform (Fig. 4.14) and Chen Keqiang for the degradation of hydroxyethylcellulose (Fig. 4.15) and polyethylene oxide (Fig. 4.16) in aqueous solution.

4.3.6 External pressure
Several workers have investigated the effect of pressure on ultrasonic degradation and obtained contradictory results. For example, whereas Schmid and Rommel observed that 5 and 8 atm respectively of excess pressure slowed and eventually stopped the ultrasonic degradation of nitrocellulose, the application of higher excess pressures, 8 and 15 atm, only resulted in a slowing down of the degradation rate for styrene, and then to the same extent for both pressures (Fig. 4.17). In contrast, Mark, although employing a lower molar mass polystyrene sample than Schmid (140 000 compared with 850 000), had observed increases in the degradation rate with increases in pressure (Fig. 4.18).

Since both authors employed intensities equivalent to a pressure amplitude (P_A) of approximately 5 atm it was anticipated that if cavitation were the cause of degradation then the extent of degradation and its rate ought to have ceased completely at 15 atm (appprox. P_h) since P_h was larger than P_A (see General Principles, Section 2.6.5).

Melville and Murray have investigated the degradation of polystyrene in toluene, albeit in vacuo and at a slightly lower insonation frequency (213 kHz) and compared their results with those obtained by Schmid (300 kHz) at 15 atm (Table 4.9, p. 122).

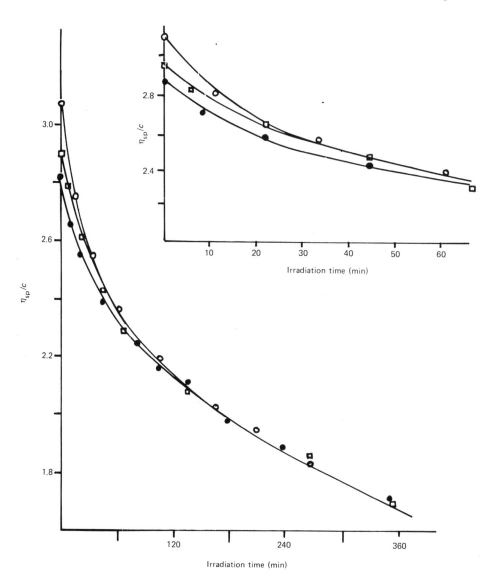

Fig. 4.10 — Effect of polymer composition (1% in benzene) on degradation rate. ○ Poly-(methyl methacrylate) — RMM, 725 000; □ Poly(methyl methacrylate)-acrylonitrile copolymer (411:1) — RMM, 615 000; ● Poly(methyl methacrylate)-polyaerylonitrile copolymer (40:1) — RMM, 544 000.

Apart from the initial values, there is a reasonable similarity between the results, i.e. initial rapid polymer breakdown merging into slower rates as the degradation proceeds. If we are to conclude that cavitation is absent in the absence of a gas, or in the presence of a large overpressure, the above data suggests that degradation can still be accomplished by ultrasonic waves even in the absence of cavitation. One

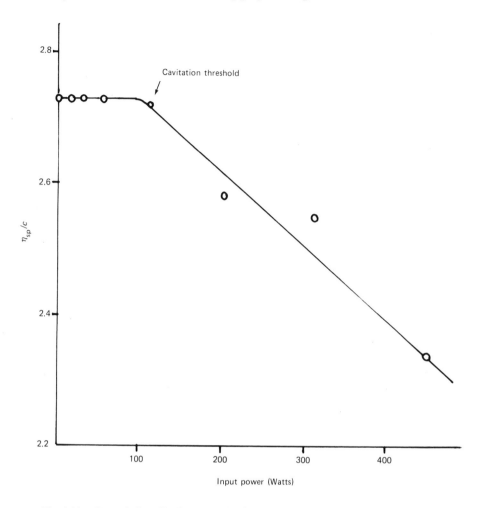

Fig. 4.11 — Degradation of hydroxycellulose (0.8%, $f = 400$ kHz) as a function of input power.

possible explanation is that shear forces are set up by the ultrasonic waves, the magnitude of which increase with increasing intensity of the wave, and it is these shear forces which are responsible for the observed degradation above. In the presence of a gas, the forces will contain an added contribution from cavitation. The values for the limiting RMM's found by Melville, Schmid and Mark are given in Table 4.l0, p. 123.

Whereas the work of Melville and Schmid seems to confirm the importance of cavitation in polymer degradation (i.e. less degradation in degassed or pressurised systems), Mark's value for the pressurised system is somewhat difficult to explain.

4.3.7 RMM and concentration
Schmid and Rommel, and separately Mostafa, have studied the degradation of a number of polystyrene samples of differing RMM. All authors found that the

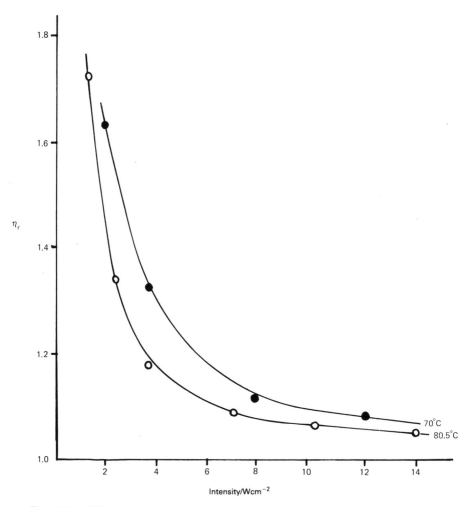

Fig. 4.12 — Effect of intensity on the degradation of polystyrene in toluene. Initial RMM, 300 000; irradiation time, 90 min.

degradation rate (Figs 4.3 and 4.4) was highest for the sample with the largest RMM and that the curves for all samples converged in the later stages, apparently reaching the same final relative molar mass.

Schmid and Rommel also found similar relationships in their investigations of polyvinyl acetate, polymethylacrylate and nitrocellulose. Basedow and Ebert [14], investigating the effect of ultrasound (20 kHz, 10 W cm^{-2}) on the degradation of dextran, observed that for molar masses (M) higher than the limiting value the degradation rate was proportional to $M^{4/3}$; for molar masses below the limiting value the rate constant was proportional to $M^{5/6}$.

Several authors have also investigated the effect of solution concentration on degradation. In every case, the rate and extent of degradation have been found to decrease with increase in concentration (Fig. 4.19).

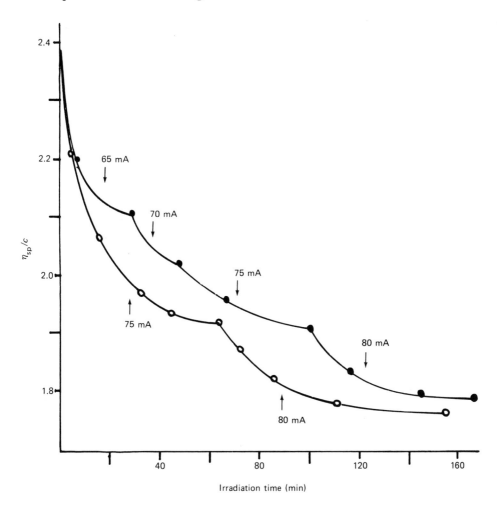

Fig. 4. 13 — Effect of intensity on the extent and rate of degradation of polystyrene in benzene.

Such observations have been interpreted in terms of the increase in viscosity of the solution — i.e. the higher the viscosity the more difficult it becomes to cavitate the solution, at a given intensity, and the smaller is the degradation effect.

The above may be summarised by Table 4.11.

In an attempt to explain these observations several models, based upon kinetic or molecular considerations, have been reported.

4.4   DEGRADATION MECHANISMS

Jellinek's model for polymer breakdown is simply an extension of the concepts identified by Schmid except that the frictional and impact forces were considered to be a consequence of cavity collapse. Using equations developed by Noltingk and

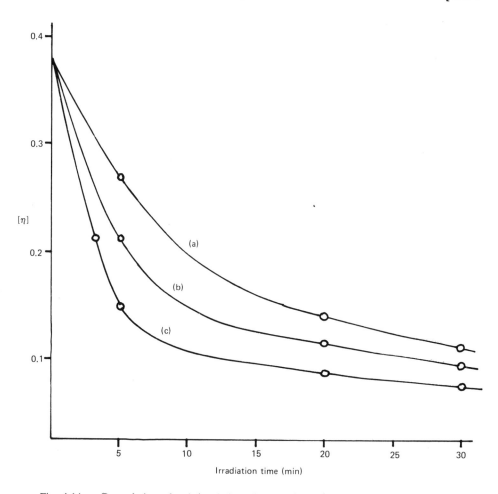

Fig. 4.14 — Degradation of poly(methyl methacrylate) in chloroform (0.1%) at various
irradiation intensities. (a) 0.8 W cm^{-2}; (b) 1.3 W cm^{-2}; (c) 2.2 W cm^{-2}.

Neppiras, Jellinek deduced that the velocity of a solvent molecule in the vicinity of a
collapsing bubble, containing a monatomic gas, was approximately 300 cm s^{-1}. By
employing a modified Stokes' formula (eqn 4.1) he was able to show that this velocity
produced a frictional force of 3.2×10^{-8} N per bond, an order of magnitude larger
than that necessary to break a C–C bond. Further, it is possible to deduce from
equation (4.1) that for a solvent velocity of this magnitude (300 cm s^{-1}) a polystyrene
molecule with a degree of polymerisation less than 430 (i.e. $M_n = 43\,000$) ought not
be degraded — i.e. there is a lower limit of RMM.

Doulah, on the other hand, has suggested that it is the shock wave energy,
released as the cavity collapses, not the shear force which is responsible for the
degradation. The suggestion is that the bubble, in the cavitation field, acts as a power
transformer converting the acoustic energy into hydrodynamic energy. By consider-

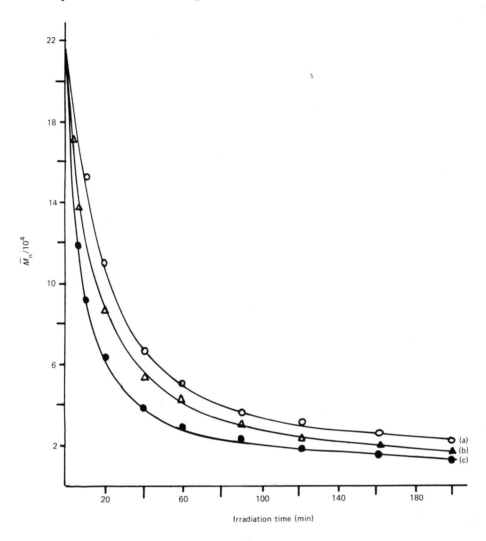

Fig. 4.15 — Degradation of an aqueous solution of hydroxyethyl cellulose (0.5%) at various sonic intensities ($T = 25°C$). (a) 1.5 A; (b) 2.0 A; (c) 2.5 A.

ing the shock waves to give rise to a series of eddies and that the macromolecule, suspended in the cavitation field, will experience motions of various amplitude and intensity due to the eddy motion, he was able to calculate the dynamic force set up across the length of the macromolecule. The results of his calculations indicated that the degradation rate was dependent upon (a) acoustic intensity (b) the size of the macromolecule, and (c) suggested that there would be a minimum chain length below which degradation would not occur. All three deductions have been confirmed experimentally. However, what neither of the above two theories could predict was the point, or points, in the macromolecule where bond breakage would

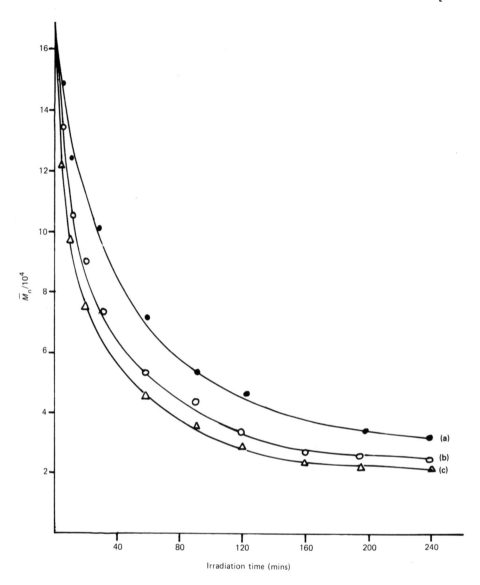

Fig. 4.16 — Degradation of aqueous polyethylene oxide (0.5%) at various sonic intensities
($T = 25°C$). (a) 1.5 A; (b) 2.0 A; (c) 2.5 A.

occur. Although it might be expected that ultrasonic degradation, like chemical or
photodegradation, occurs at the points of inherent weakness in the polymer's
backbone, work by Melville seems to refute this suggestion. For example, Melville
has shown that the degradation of two separate poly(methyl methacrylate)(A)–
polyacrylonitrile (B) copolymers (molar ratios of A:B of approximately 40:1 and
400:1 respectively) yielded almost identical final molar masses, as determined by
osmometry, after irradiation with ultrasound (Table 4.12, p. 124).

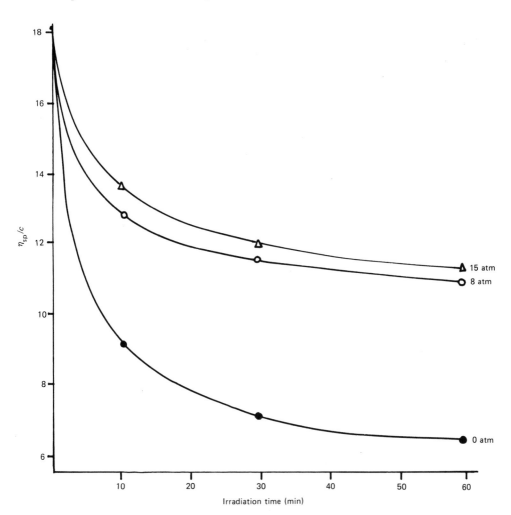

Fig. 4.17 — Degradation of polystyrene in toluene at various applied pressures △ 15 atm of applied pressure; ○ 8 atm of applied pressure; ● 0 atm of applied pressure.

If it is assumed that the A–B linkages are appreciably weaker than the corresponding A–A or B–B linkages, then appreciably different molar masses ought to have been obtained as is the case for thermal degradation. For example we might expect that the RMM would fall to the average value of the methyl methacrylate chains which for the 40:1 copolymer (initial RMM 547 000) would be 4000 (= 40 × 101), while for the copolymer of RMM 615 000 this would be 41 000 (= 411 × 101). Further confirmation that degradation is not taking place selectively at the A–B linkages is afforded by Fig. 4.10 where it can be seen there is no appreciable difference in the rate of degradation of the two copolymers.

If scission does not take place at the weak spots within the macromolecule we are

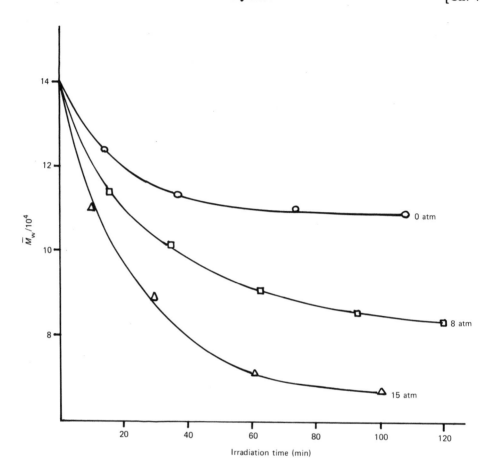

Fig. 4. 18 — Degradation of polystyrene in toluene at various applied pressures. $T = 20°C$; $f = 300$ kHz; intensity = 10 W cm^{-2}.

Table 4.9 — Degradation of polystyrene in toluene

Irridation time (min)	Melville (vacuo) η_{sp}/c	Schmid (15 atm) η_{sp}/c
0	2.72	15.7
5	2.38	2.5
20	2.03	2.0
60	1.41	1.8
100	1.14	—
120	—	1.6

Table 4.10 — Limiting RMM for polystyrene in toluene

RMM	Melville		Schmid		Mark	
Initial	700 000		850 000		140 000	
	Vacuo	1 atm	15 atm	1 atm	15 atm	1 atm
Final	226 000	47 000	121 000	28 000	70 000	110 000

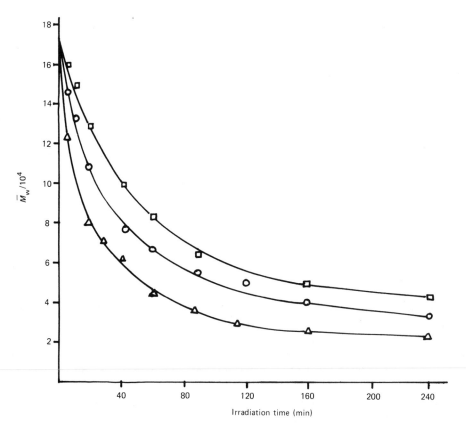

Fig. 4.19 — Degradation of aqueous poly(ethylene oxide) at various concentrations. (Intensity = 2.5 A; temperature = 25°C). □ = 1.5%; ○ = 1.0%; △ = 0.5%.

left to conclude that scission is either a random process or that it takes place at the midpoint of the polymer chains. Elegant work by Glynn, Van der Hoff and Reilly [15] seems to suggest that it is the latter, or more correctly there is a gaussian distribution about the midpoint of a chain. However, as to whether the main chains

Table 4.11 — Factors which increase the rate and extent of degradation

Saturation of the polymer solution with a gas.
Employing a gas with a low solubility.
Using a solvent with a low vapour pressure.
Reducing the experimental temperature.
Reducing the insonation frequency.
Increasing the intensity of irradiation.
Decreasing the solution concentration.
Increasing the RMM of the polymer.

Table 4.12 — Comparison of effect of irradiation (~ 6 h) on RMM of various polymers

Polymer/copolymer	Initial RMM	Final RMM
Methyl methacrylate	725 000	544 000 ± 40 000
Methyl methacrylate*	725 000	460 000 ± 40 000
Methyl methacrylate- acrylonitrile	615 000	385 000 ± 25 000
Methyl methacrylate- acrylonitrile*	615 000	395 000 ± 25 000

* Repeat determination of RMM.

are primarily broken by ultrasonic action is still open to question since it is possible that the main chain scissions are secondary effects due to chemical reactions initiated by unstable intermediates, such as free radicals or ions, produced by the ultrasound. For example Ramsden and McKay [16] have shown that hydroxyl radicals, generated by the oxidation of Fe^{2+} by H_2O_2, are the cause of chain scission in polyacrylamide molecules in aqueous solution.

Whatever the mechanism involved, depolymerisation results in the breakage of an atomic bond in the macromolecule to produce long-chain radical entities. The existence of these radical entities has been established by investigating the depolymerisation process in the presence of various radical scavengers such as diphenyl picrylhydracyl (DPPH). Spectroscopic analysis of the loss of DPPH ($\lambda_{max} = 525$ nm) has allowed both a determination of the depolymerisation rate and the number of break points (i.e. 2 DPPH molecules per bond breakage). Evidence for the formation of macroradicals in the degradation of poly(methyl methacrylate), polystyrene and poly(vinyl acetate) has also been provided by Tabata and Sohma [17] using spin trapping and e.s.r. techniques.

Obviously in the absence of radical scavengers the molecular fragments are free to recombine by the usual termination mechanisms. Combination termination will produce macromolecules of the same, or differing lengths, as those existing just prior to bond breakage, depending upon whether the combining molecular fragments are from the same or differing polymer chains. Disproportionation termination between fragments no matter what their origin must result in smaller macromolecules. Thus in the long term it might be anticipated that degradation will take place with a decrease in RMM. What perhaps is not expected is the observed reduction in polydispersity (i.e. ratio of $\bar{M}_w$:$\bar{M}_n$) with the passage of time (Table 4.13).

Table 4.13 — Ultrasonic degradation (20 kHz, 70 W, 27°C) of various polyacrylates (PRMA) in toluene: degradation rate versus gas solubility

Irradiation time/min	Heterogeneity index $\bar{M}_w/\bar{M}_n$			
	R = Methyl	Ethyl	n-Butyl	Cyclohexyl
0	4.20	2.80	2.66	2.40
60	2.07	1.80	1.80	1.68
120	1.77	1.64	1.72	1.58
180	1.58	1.62	1.90	1.50
240	1.46	1.48	1.80	1.49
360	1.45	1.54	1.66	1.46
480	1.38	1.43	1.62	1.45

This reduction has significant consequences, since although the absolute magnitude of a polymer's relative molar mass ($\bar{M}_w$ or $\bar{M}_n$) is important, there is a whole series of physicochemical properties (e.g. film formation, chemical stability, solution flow, etc.) which depend upon the degree of polydispersion. With hindsight the reduction in polydispersity can be explained using the ideas of Glynn and Jellinek. For example, according to Jellinek the larger the polymer the greater is the expected frictional force and the more likely that bond breakage will occur. If scission is gaussian (Glynn), then breakage will produce molecular fractions of slightly different size and if the larger fractions, from two separate chains, recombine the resultant polymer will still be sufficiently large to be further degraded. Combination of the smaller fractions, however, will lead to a decrease in molecular size and the chains will have a lower probability of further degradation. With the passage of time there will be a higher proportion of the smaller RMM material in the system and the sample will become more homogeneous. Although most of the reported literature seems to infer that degradation of a macromolecule takes place by main-chain scission, we should not rule out the possibility of C–H scission since the bond energy for the latter (412 kJ mol^{-1}) is only some 20% higher than that for the former (350 kJ mol^{-1}). The consequence of such a scission would, on termination by combination, lead to branched structures. Interestingly, our own studies have revealed that both the shape of the macromolecule and the glass transition temperature, T_g, change with irradiation time.

It is well known that the structure, physical and physicochemical characteristics of macromolecules depend not only on the nature of the monomer (homopolymer) or monomers (copolymer) from which they are synthesised but also on the method of production (e.g. graft, block or random copolymers). For instance, whereas random copolymerisation of two monomers A and B produces a polymer with a property (e.g. solubility, polarity) which is the weighted average of the two constituent monomers, block (or graft) copolymers have properties which are the sum of those of the two homopolymers. For example a graft copolymer consisting of water soluble (poly(ethylene oxide)) and oil soluble (polystyrene) components is capable of dissolving in both solvent types. The preparation of graft copolymers can be accomplished by irradiating a mixture of two homopolymers with either γ or X-rays, or by polymerisation of a monomer (A) from initiation sites along the polymer chain of monomer B. Alternatively the copolymer can be prepared by a mechanical blending of the two homopolymers. To effect the latter type of process and produce a genuine graft copolymer rather than a simple mixture of the two homopolymers the two components are milled or masticated together. The mechanical shearing during the milling process results in homolytic cleavage of the bonds and subsequent cross combination between the different polymer fragments then results in a graft copolymer. Block copolymers on the other hand are often produced via anionic mechanisms. The process utilises the fact that the anionic chain ends often remain 'alive', particularly at low temperature, even though the monomer has been consumed and that chain growth can be restarted by addition of more of the same monomer, to give a homopolymer, or addition of a different monomer to give the block copolymer. In practice some restrictions exist with respect to the type of monomers which can be used in a block copolymerisation reaction in that the monomers should have similar electron affinities if mutual reinitiation is to take place. However as with graft copolymer production, milling of a mixture of homopolymers can also be used to prepare the block copolymer. This technique alleviates many of the experimental difficulties which may be experienced using the anionic technique.

We have already seen in this section how ultrasound, like milling, is capable of producing chain scission and thus give rise to macroradicals. It is therefore not surprising that the application of ultrasound to the synthesis and modification of polymers has attracted, and continues to attract, the attention of many investigators.

4.5 POLYMERISATION

Until recently most investigations involved applying ultrasound to systems containing either a mixture of the homopolymers or a mixture of polymer and monomer in the hope of producing graft or block copolymers. The use of ultrasound to initiate the polymerisation of systems containing only monomer, or monomer plus initiator has received little attention. This no doubt is due to the fact such systems suffer from the complication that as the polymerisation proceeds and the concentration of polymer increases, the competing depolymerisation reaction will become increasingly more

significant. It is also likely that the increase in viscosity accompanying the polymerisation reaction will lead to a change in the acoustic environment within the system.

Keqiang has successfully produced block copolymers, based upon cellulose,
while Henglein has produced both graft and block copolymers using polystyrene and
poly(methyl methacrylate). Malhorta [18] has employed a variety of homopolymers
(rigid and flexible) but has met with only limited success in the synthesis of block
copolymers. Homopolymers with bulky substituents, although able to undergo chain
scission did not result in scrambled copolymers in the presence of polystyrene. In all
of the above syntheses, however, degradation of the homopolymers by ultrasound
must first provide long-chain radicals of each component which can then terminate
by cross combination. Obviously to ensure production of the macroradical both
homopolymers must have degrees of polymerisation greater than P_L. We have
already seen (Table 4.2) that polystyrene samples of approximately 200 000 are
degraded to give polymers with final RMM values of approximately 70000. If
Glynn's gaussian distribution mechanism is correct this implies only two C–C bonds
are broken overall, i.e. 200 000 to 100 000 and then 100 000 to 50 000. Similar
deductions can be made from Tables 4.8 and 4.10. However this is an overall
resultant value for chain scission and does not take into account that bonds are being
remade due to combination termination. Malhorta has deduced from GPC analysis
that the rate of bond scission in a polystyrene sample (2% in THF), at 27°C, after 30
min irradiation (20 kHz, 70 W) is 0.032 scissions per minute (i.e. 1 bond in 30 min)
while after 8 hours it is 0.014 scissions per minute — i.e. a total of 7. At a lower
temperature (-20°C) the number is increased considerably, as expected, being
0.116 scissions per minute after 30 min and 0.053 scissions per minute after 4 h. What
may be surprising therefore is the high yield of block copolymers produced, often as
high as 90%, which results from the breakage of only a limited number of bonds in
the macromolecule. This has prompted Berlin [19] to suggest that the macroradical is
also capable of degrading a stable macromolecule, a suggestion which is supported
by the work of Ramsden and McKay. They reported that hydroxyl radicals were
capable of inducing chemical degradation of polyacrylamide (see above).

Block copolymers have also been produced by irradiating a solution containing a
homopolymer (from monomer type A) and a monomer (type B). In such cases
polymerisation of the monomer (B) is thought to be initiated by the macroradical
produced by the ultrasonic degradation of the homopolymer since in cases where
polymer was absent polymerisation was not observed to occur. Similar findings have
been reported by other workers when attempting to homopolymerise pure vinyl
monomers in the presence of ultrasound. However these observations are in direct
contrast to the findings of other workers including ourselves. Ultrasonic waves ($I = 8$
W cm^{-2}) have been found to initiate the polymerisation of acrylonitrile in aqueous
media saturated with N_2. The initiating species is presumably HO· radicals from the
decomposition of water. Berlin confirms this opinion in his investigation of the
polymerisation of polystyrene in the presence of styrene monomer since addition of
water to the solvent (benzene) greatly enhanced the yield of polymer. It could be
argued, however, that the appearance of water decomposition products (e.g. H_2O_2)
led to oxidation of the various impurities, which previously, may have acted as
inhibitors.

Melville has studied the ultrasonically induced polymerisation of monomers such as styrene, methylmethacrylate and vinyl acetate in the presence and absence of poly(methylmethacrylate). In that the polymerisation rates ($\sim 1\%$ conversion/h) were not substantially increased by the presence of polymer, he concluded that the degradation of polymer was not the major source of radical production. Also using hydroquinone as an inhibitor he was able to deduce from retardation times that the rate of radical production was $\sim 2 \times 10^{-9}$ mol dm^3 s^{-1}. A typical value for radical production using the thermal initiation of AZBN say (10^{-3} mol dm^3) at 60°C is $\sim 2 \times 10^{-8}$ mol dm^3 s^{-1} — i.e. an order of magnitude greater.

Ultrasonic waves have also been found to increase the rates of emulsion and suspension polymerisations. For example, we have found a twofold increase in the thermally initiated production of polystyrene by emulsion polymerisation when in the presence of ultrasound (20 kHz). Possible explanations for these increases include:

(i) the oxidation of impurities (see above),
(ii) the removal of oxygen (known to inhibit radical reactions) by ultrasonic degassing, and
(iii) ultrasonic degradation of the polymer to provide more active sites (i.e. autocatalysis).

Hatate *et al.* [20] have investigated the suspension polymerisation of styrene under ultrasonic irradiation in both batch and continuously stirred reactors. For both processes they were able to confirm that irradiation was an effective method of both preventing agglomeration between the droplets and the sticking of droplets to the reactor wall, both of which lead to serious problems such as heat build-up and the formation of large masses and make it impossible to carry out prolonged operations. In the case of the batch reactor, the effects of ultrasound on the conversion and the average RMM were found to be negligibly small. Hatate *et al.* suggested that the ultrasonic energy supplied to the monomer phase (500 W), while sufficient to prevent agglomeration was not great enough to affect the polymerisation rate. Unfortunately it was not possible to quantify the effect of ultrasound in the continuous process due to the extreme difficulty in the absence of ultrasound in realising the steady state owing to agglomeration and sticking of the droplets.

Few workers have investigated the effects of varying such parameters as frequency, intensity, temperature and the nature of the gas on the polymerisation process. Berlin has shown that for the block copolymerisation of poly(methyl methacrylate) with acrylonitrile, the time required to produce a given amount of polyacrylonitrile in the block decreased with increasing intensity. Kruus, Donaldson and Farrington [21] have been able to show that in the absence of a thermal initiator (e.g. AZBN) there is a propagation rate (R_p) dependence for the bulk polymerisation of methyl methacrylate on the square root of the sound intensity ($I^{1/2}$). This is similar to that observed for photo-initiated polymerisations where the propagation rate is proportional to the square root of the light intensity.

We have found that at a given temperature and irradiation frequency ($T = 60$,

$f = 20$ kHz), but in the presence of AZBN, there is a maximum in the R_p versus I curve for the solution polymerisation of NVC. At very high intensities (> 100 W cm^{-2}) the conversion was negligible. We have also observed that at a given power there is an optimum irradiation time beyond which there is a decrease in the amount of polymer produced and that at lower temperatures and intensities the optimum polymer yields are increased (Fig 4.20).

Driscoll and Sridhari [22] have also obtained, as predicted (see 2.6.3), an inverse temperature dependence of polymerisation rate. For example, for the polymerisation of styrene and methylmethacrylate in the presence of their respective homopolymers they observed that the lower was the reaction temperature the faster was the reaction rate and the higher the final polymer yield (Figs 4.21 and 4.22).

We [23] have also investigated the effect of ultrasound on the aqueous polymerisation of N-vinyl pyrrolidinone (NVP). This particular monomer does not follow the normal rate (R_p)-monomer (M) dependence ($R_p = K[M]$, Fig. 4.23a), but exhibits a maximum in the rate at 80% monomer (v/v) of monomer to water.

This maximum is thought, from viscosity and enthalpy of mixing studies (Fig. 4.24), to be due to the formation of an H-bonded monomer–water complex. Our previous investigations of the solvolysis of 2-chloro-2-methylpropane had led us to conclude that low intensity ultrasound (< 2 W cm^{-2}) was capable of destroying H-bonds and if this were so then application of ultrasound to the NVP system ought to lead to a destruction of the complex and a decrease in R_p. Except for the pure monomer system (where there could not be H-bonding since water was absent) all R_p values were decreased in the presence of ultrasound (Fig. 4.23b). It can only be assumed that some other effect, perhaps chemical rather than physical, is operating in the pure monomer system since the R_p value increased in the presence of ultrasound.

Henglein [24] has shown that the degree of polymerisation decreases when the duration of pulsed ultrasound is decreased. The growth and collapse of cavitation bubbles require a finite time — they are not instantaneous. Henglein has also shown that the rate (and degree of polymerisation) depends upon the nature of the gas used to saturate the system. For the polymerisation of methacrylic acid in aqueous solution, a 15-minute irradiation yielded 10.7% conversion in the presence of N_2, 1.8% conversion in the presence of O_2 (low presumably due to inhibition) and no polymerisation in a degassed solution).

Miyata and Nakashio [25] have studied the effect of frequency and intensity on the bulk polymerisation of styrene. They found that whereas the mechanism of polymerisation was not affected by the ultrasound, the overall rate constant, k ($R_p = k[M]$) decreased linearly with increase in the intensity whilst the average RMM increased slightly. The decrease in the overall value of k they interpreted as being caused by either an increase in the termination reaction (i.e. specifically the termination rate constant, k_t) or a decrease in the initiator efficiency. The increase in k_t ($= k_t^o/\eta$) is the more reasonable in that ultrasound is known to reduce the viscosity of polymer solutions. (This reduction in viscosity and consequent increase in k_t could account for our observed reductions in rate in the NVP system. However this explanation does not account for the large rate increase observed for the pure monomer system.) Unfortunately little can be learned from Miyata's and Nakashio's

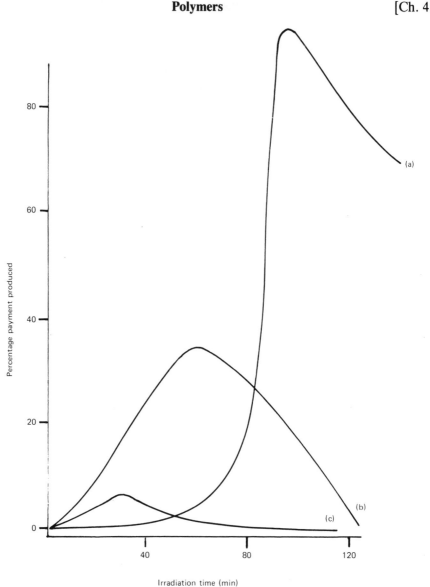

Fig. 4.20 — Effect of ultrasonic intensity and temperature on the thermally initiated polymerisation of N-vinyl carbazole. (a) $T = 60°C$; intensity $= 27.6$ W cm^{-2}; (b) $T = 50°C$; intensity $=$
59.6 W cm^{-2}; (c) $T = 60°C$; intensity $= 59.6$ W cm^{-2}.

work at different frequencies. Employing a constant output power from the sound
generator, and operating at 200, 400, 600 and 800 kHz they observed a maximum for
the value of k at 400 kHz. Whether the maximum is meaningful is uncertain because
of the experimental conditions employed. It has already been suggested that larger
input intensities are necessary, at the higher frequencies, if equivalent sonochemical
effects are to be produced.

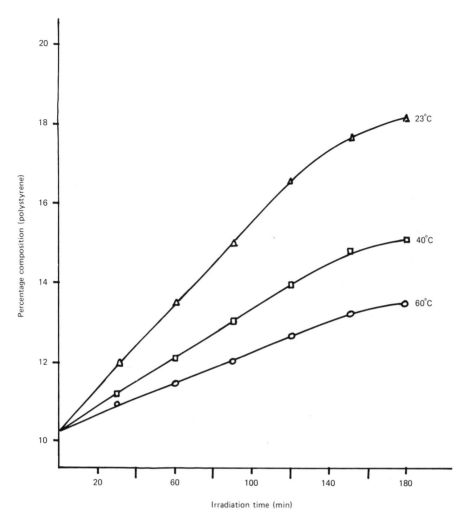

Fig. 4.21 — Continuous polymerisation of styrene in the presence of polystyrene ($\bar{M}_n = 83\,200$) using 70 W ultrasound ($f = 23$ kHz). Reactor feed: 10.1 wt% polystyrene; 40% toluene; 10% acetone; 4.52 mol dm^{-3} styrene.

Kruus is one of the few investigators who has attempted to interpret quantitatively the results of ultrasonically induced polymerisations. For the polymerisation of nitrobenzene he has observed that the polymerisation rate (measured as solution darkening) decreased both in the presence of gases with high solubility (e.g. SO_2 and CO_2) and solutes with high vapour pressures. By assuming that the species responsible for initiating the polymerisation was a radical and that it was produced in the cavitation bubble, he was able to deduce the lower and upper limits for the temperature of the bubble as 600 K and 40 000 K respectively. Using a similar mechanistic model he has been able to explain the conversion of methyl methacrylate (bulk polymerisation, no initiator, 20 kHz) in terms of reaction time, reaction

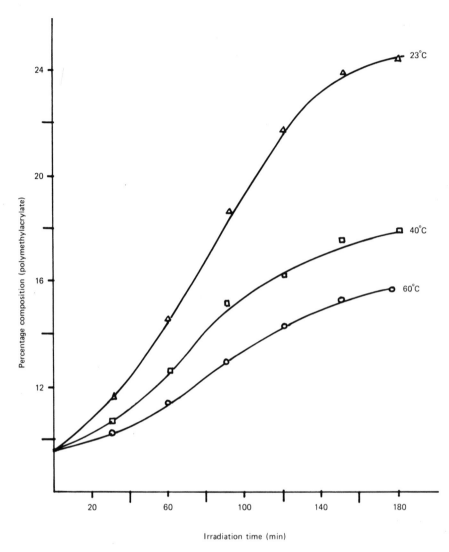

Fig. 4.22 — Continuous polymerisation of methyl methacrylate in the presence of poly(methyl methacrylate) ($\bar{M}_n = 124\,900$) using 70 W ultrasound ($f = 23$ kHz). Reactor feed: 9.65 wt% polymer; 4.7 mol dm^{-3} monomer; 40% toluene; 10% acetone.

volume and ultrasonic intensity — Fig. 4.25. The conversion (3% per hour) and RMM (MW = 700000) of the resultant polymer ($T = 40°C$, $I = 20$ W cm^{-2}) was considerably lower than that found in our own laboratories for the ultrasonically initiated solution polymerisation of NVC (Yield = 60%/h, $T = 60°C$; $M_w > 2 \times 10^6$). Kruus has also determined activation energies for the bulk polymerisation of methyl methacrylate in the presence of ultrasound. The value obtained (~ 19 kJ mol^{-1}) was similar to that observed for the bulk thermal polymerisation reaction (~ 17–20 kJ mol^{-1}), provided the contribution from the initiation step was excluded. This close

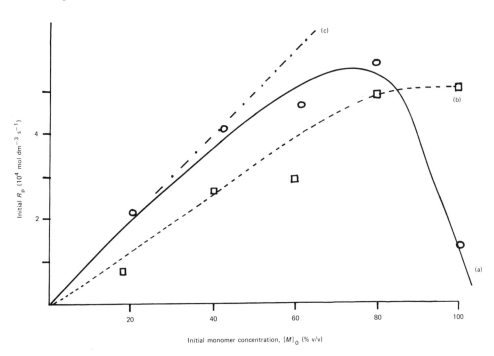

Fig. 4.23 — Effect of ultrasound on the aqueous polymerisation of N-vinyl pyrrolidinone. (a) R_p versus $[M]_0$ dependence without ultrasound; (b) R_p versus $[M]_0$ dependence with ultrasound; (c)R_p versus $[M]_o$ dependence theoretical.

correspondence in activation energies suggests that the effective activation energy for the initiation step, in the presence of ultrasound, may be taken to be 0 kJ mol^{-1}, as is the case in photopolymerisation. Kruus also studied the polymerisation in the presence of the radical scavenger DPPH. Unfortunately the rates of initiation as deduced from both the overall polymerisation rate and the rate of DPPH consumption, although of similar magnitude, showed inexplicable differences in temperature dependence.

Toppare, Eren and Akbulut [26] have investigated the effect of ultrasound (25 kHz bath) on both the polymerisation rate and composition of the copolymers produced by the electro-initiated cationic polymerisation of isoprene with α-methyl-styrene. In the absence of ultrasound the yield of copolymer was found to decrease with increase in the applied polymerisation potential due to the formation at the electrode of a polymer film. These films created a resistance to the passage of current into the bulk media with consequent reduction in rate and yield. In the presence of ultrasound however, the total conversion was found to exhibit a slight increase (Fig. 4.26) with increase in the polymerisation potential (E_{pol}). This increase was attributed to a 'sweeping clean' of the electrode surface by the ultrasound. The authors also noted that both the proportion of isoprene incorporated into the polymer (Fig. 4.27) and the reactivity ratios (Table 4.14, p. 138) were affected by the polymerisation potential.

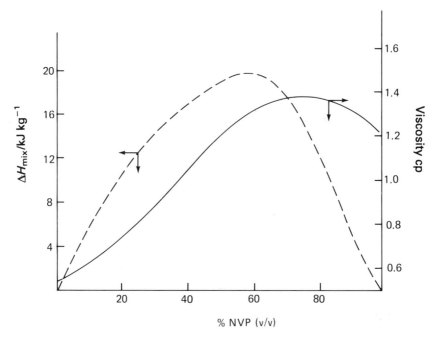

Fig. 4.24 — Heat of mixing (———) and viscosity (– – –) for aqueous *N*-vinyl pyrrolidinone mixtures.

Apart from that of Miyata and Nakashio little work has been performed on the effect of frequency on polymerisation rate (and yield). Frequency is a factor which determines cavitation threshold, size of bubble and the time scale of bubble growth and collapse. It is anticipated that changing frequency will alter the polymerisation process.

REFERENCES

[1] G. Schmid and O. Rommel, *Z. Phys. Chem.*, 1939, **A185**, 97; *Elektrochem.*, 1939, **45** 659. G. Schmid and E. Beuttenmuller, *Elektrochem.*, 1943, **49**, 325; 1944, **50**, 209. G. Schmid, *Phys. Z.*, 1940, **41**, 326; *Z. Phys. Chem.*, 1940, **186A**, 113. G. Schmid, P. Paret and H. Pfleider, *Kolloidn. Zh.*, 1951, **124**, 150.

[2] H. F. Mark, *J. Acoust. Soc. Am.*, 1945, **16**, 183.

[3] W. Gaertner, *J. Acoust. Soc. Am.*, 1954, **26**, 977.

[4] R. O. Prudhomme and P. Graber, *J. Chim. Phys.*, 1949, **46**, 667. R. O. Prudhomme, *J. Chim. Phys.*, 1950, **47**, 795.

[5] H. H. Jellinek and G. White, *J. Poly. Sci.*, 1951, **6**, 745; 1951, **7**, 33; 1954, **13**, 441.

[6] A. Weissler, *J. Appl. Phys.*, 1950, **21**, 171; *J. Chem. Phys.*, 1950, **18**, 1513; *J. Acoust. Soc. Am.*, 1951, **23**, 370.

[7] G. Gooberman, *J. Poly. Sci.*, 1960, **42**, 25. G. Gooberman and J. Lamb, *J. Poly Sci.*, 1960, **42**, 35.

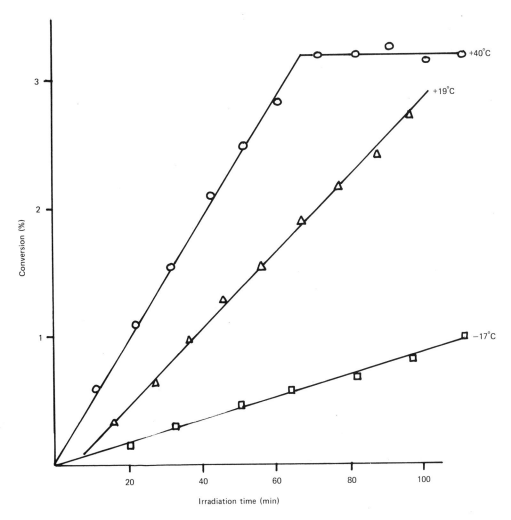

Fig. 4. 25 — Bulk polymerisation of methyl methacrylate in the presence of ultrasound. (Intensity = 20 W cm^{-2}; f = 20 kHz). ○, + 40°C; △, + 19°C; □, − 17°C.

[8] F. Gebert, *Angew. Chem.* 1952, **64**, 625.

[9] H. W. Melville and A. J. R. Murray, *J. Chem. Soc. Faraday. Trans.*, 1950, **46**, 996.

[10] M. S. Doulah, *J. Appl. Poly. Sci.* 1978, **22**, 1735.

[11] M. A. K. Mostafa, *J. Poly. Sci.*, 1958, **33**, 295; 1958, **33**, 311, 1958, **33**, 323.

[12] E. Wada and H. Nakane, *J. Sci. Research Inst.* (*Tokyo*), 45, **1**, 1951.

[13] C. Keqiang, S. Ye, L. Huilin and X.Xi, *J. Macromol. Sci. Chem.*, 1985, **A22**, 455; 1986, **A23**, 1415.

[14] A. Basedow and K. H. Ebert, *Makromol.Chem.*, 1975, **176**, 745; *Polymer Bulletin*, 1979, **1**, 299.

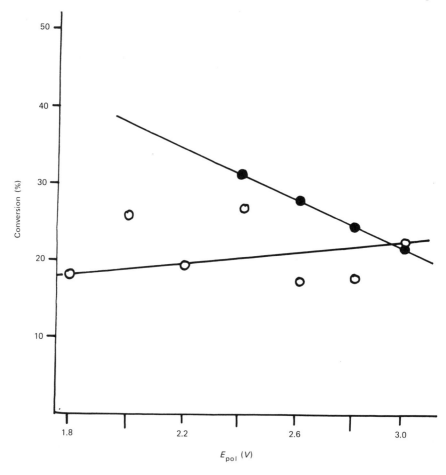

Fig. 4.26 — Effect of polymerisation potential (E_{pol}) on the percentage conversion in electro initiated copolymerisation of isoprene with α-methyl styrene. ● Without ultrasound; ○with ultrasound.

[15] P. A. R. Glynn, B. M. E. Van der Hoff and P. M. Reilly, *J. Macromol. Sci.*, 1972, **A6**, 1653; 1973, **A7**, 1695.

[16] D. K. Ramsden and K. McKay, *Polym. Deg. Stab.*, 1986, **15**, 15.

[17] M. Tabata and J. Sohma, *Chem. Phys. Lett.*, 1980, **73**, 178; *Eur. Polym. J.*, 1980, **16**, 589.

[18] S. L. Malhorta, *J. Macromol. Sci. Chem.*, 1982, **A18**, 1055.

[19] A. A. Berlin, *Usp. Khim.* 1960, **29**, 1189; *Khim. Nauka. Prom.*, 1957, **2**, 667.

[20] Y. Hatate, T. Ikeura, M. Shinonome, K. Kondo and F. Nakashio, *J. Chem. Eng. Jpn.*, 1981, **4**, 38.

[21] P. Kruus, D. J. Donaldson and M. D. Farrington, *J. Phys. Chem.*, 1979, **83**, 3130.

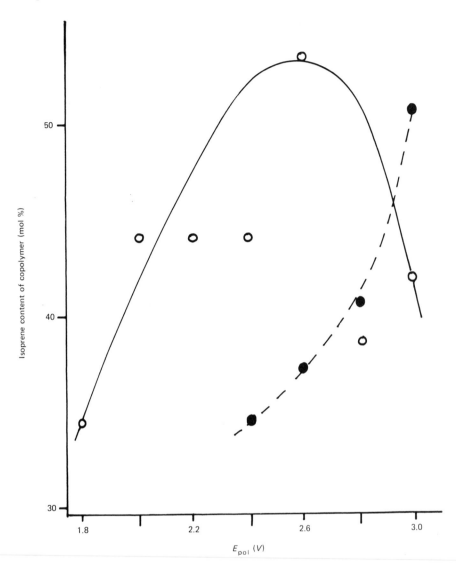

Fig. 4.27 — Effect of polymerisation potential (E_{pol}) on the isoprene content (mol%) of the electrochemically obtained isoprene-α-methyl styrene copolymers. ○ With ultrasound; ● without ultrasound.

[22] K. F. Driscoll and A. U. Sridhari, *J. Appl. Polym. Sci., Appl. Polym. Symp.*, 1975, **26**, 135.

[23] J. P. Lorimer and T. J. Mason, *Ultrasonics International* 87, *Conference Proceedings*, p. 762, Butterworths.

[24] A.Henglein, *Z. Naturforsch.*, B, 1952, **7,** 484; B, 1955, **10**, 616; *Makromol. Chem.*, 1954, **14,**15; 1955, **15**, 188; 1956, **18**, 37.

Table 4.14 — Effect of polymerisation potential on reactivity ratio

Epol/V	Reactivity ratios		[η]
	Isoprene	α-methylstyrene	
2.4	0.26(0.81)	0.19(0.78)	0.112(0.049)
2.6	0.36(0.77)	0.23(0.77)	0.110(0.057)
2.8	0.48(0.70)	0.50(0.98)	0.098(0.045)
3.0	0.80(0.97)	0.75(0.97)	0.044(0.050)

Figures in parentheses are in the presence of ultrasound.

[25] T. Miyata and F. Nakashio, *J. Chem. Eng. Jpn.*, 1975, **8**, 463.
[26] L. Toppare, S. Eren and U. Akbulut, *Polymer Communications*, 1987, **28**, 36; U. Akbulut, L. Toppare and B. Yurttas, *Polymer*, 1986, **27**, 803; *British Polymer Journal*, 1986, **18**, 273.

5

Kinetics and mechanisms

For many chemical reactions the application of high power ultrasound has led to substantial improvements in both reaction rate and yield. Since the mechanistic aspects of this topic has already been covered in great detail for reactions taking place in organic media (Chapter 3), this section will be restricted mainly to a discussion of mechanisms occurring in aqueous solution. A discussion of the various methods of interpreting kinetic data in both aqueous and non-aqueous will also be included.

5.1 INTRODUCTION

In general, reactions may be divided into two groups:

(1) Those which take place normally but are *accelerated* by ultrasound, and
(2) Those which would not otherwise take place in the absence of ultrasound.

A typical example of a reaction accelerated by ultrasound is ester hydrolysis and, depending upon the type of ester involved, a range of rate enhancements have been observed. For example whereas dimethyl sulphate is hydrolysed five times faster in the presence of ultrasound, the 4-nitrophenyl esters of various carboxylic acids [1] have much smaller (< 20%) rate enhancements (see later). Similar (low) enhancements (4–14%) have also been observed for methyl ethanoate in aqueous and aqueous acetone solution [2–4]. Other reactions found to be accelerated by ultrasound include the hydration of acetylene to ethanal and the reaction of zinc or calcium carbonate with acids [5]. As a general rule, ultrasonic acceleration often occurs when one of the products is gaseous. Typical examples of enhanced gaseous reactions are the decomposition of diazocompounds [6] and the aqueous decomposition of potassium persulphate [7]. This latter reaction (9.7% rate enhancement at 8.7 kHz) may be influential in leading to the rate increases observed in ultrasonically accelerated emulsion polymerisation. One possible reason why ultrasound is effective in gaseous reactions is that it provides more effective dispersion of the gas thereby preventing the build-up of gas layers which would normally inhibit the reaction. The corollary of such a proposed mechanism would explain the increased rates, and higher current densities, achieved in electrochemical processes where polarisation at the electrode surfaces considerably influences the process itself (see Chapter 8). Other reactions benefiting from the use of ultrasound are those employing solid catalysts, where irradiation produces fine dispersions of the solid and provides for greater surface area and reactivity. Again, these have already been discussed at length in Chapter 3 and the reader is referred to that section for more detail.

Reactions which fall into the second category, i.e. those which would not normally take place without ultrasound, are depolymerisation (Chapter 4); block and graft copolymerisation (Chapter 4); the production of H_2O_2, HNO_2 and HNO_3 by the irradiation of air-saturated water; the production of hydroxyaromatic compounds by the irradiation of an aqueous solution of the aromatic; and the liberation of I_2 from aqueous KI in the presence of CCl_4. This latter reaction is quite important sonochemically since it is often used both as a demonstration experiment (i.e. the liberated I_2 can turn starch the characteristic blue colour) and as a means of measuring cavitation intensity.

The precise nature and origin of these reactions and the method by which the enhancement of rate occurs on irradiation of the solutions are factors which have not been fully resolved. However after numerous investigations the conclusion is that they are the result of cavitation and as such must be the consequence of one or other (or possibly a combination) of the following:

(a) reaction in the cavitation bubble within which there are very high temperatures and pressures;
(b) reaction as a result of secondary reactions taking place at the gas–liquid interface of the bubbles;
(c) reaction as a result of the enormous pressures released on bubble collapse.

Of the many experimental studies into the effect of ultrasound on chemical reactivity, few have received any detailed kinetic study of the effect of variations in irradiation frequency or intensity, the type of gas in the system or its concentration, or the type of solvent. Most have been investigations of the oxidation-reduction reactions of aqueous solution and have been concerned with mechanistic rather than kinetic aspects. In an attempt to determine the precise origin of sonochemical reactivity, many of the early workers adopted mechanisms similar to those known to occur in radiation chemistry. For example it is well known that when water is irradiated with ionising radiation the initial species generated are hydrated electrons, hydrogen atoms and hydroxyl radicals. Combination of these radicals, initially distributed inhomogeneously in solution, leads to the molecular species H_2 and H_2O_2 (eqns (5.1) and (5.2)) — i.e. the same products as observed when sonolysing water. Whether on sonolysis the free radicals which are generated recombine in the collapsing bubble, at the interface of the gas bubble or in the surrounding liquid is still open to question. What is accepted is that dissociation (eqn 5.3) into the primary H· and ·OH radicals takes place in the cavitation bubble where there exist very high temperatures. In fact recent investigations by Henglein and separately by Suslick suggest that there is a closer analogy between sonolysis and pyrolysis than sonochemistry and radiation chemistry.

$$H\cdot + \cdot H \rightarrow H_2 \tag{5.1}$$

$$HO\cdot + \cdot OH \rightarrow H_2O_2 \tag{5.2}$$

$$H_2O \rightarrow H\cdot + \cdot OH \tag{5.3}$$

5.2 MECHANISTIC ASPECTS

In 1950 Miller [8] studied the oxidation of air-saturated Fe^{2+} ions to Fe^{3+} ions using a frequency of 500 kHz. The Fe^{3+} yield was found to be dependent upon the length of insonation (up to 10-minutes duration), beyond which time non-linearity of yield was observed. If however at the end of the 10 minute insonation period the solution was resaturated with gas, and then further irradiated, a linear concentration change–time curve of identical slope to the first was obtained. This observation suggests that the non-linearity is due to degassing of the solution, a suggestion which may help explain the rapid initial depolymerisation observed when polymers are subject to ultrasound (Chapter 4). Most present workers in the field now usually continually bubble gas (argon) through the system when studying the effects of ultrasound. However there are a few workers whose preference is to pre-irradiate (i.e. degas slightly) before commencing an investigation.

A second experimental feature observed by Miller was that for Fe^{2+} concentrations greater than 5×10^{-4} M, the yield of Fe^{3+} ion, (measured as the loss of Fe^{2+} per minute), was independent of the initial Fe^{2+} concentration (Fig. 5.1).

Miller concluded that, similar to the effects of ionising radiation, the oxidation action was an indirect process, resulting from reaction of the Fe^{2+} with the reactive $OH\cdot$radical fragments (eqn 5.4) produced from water. If it was a direct oxidative process, independent of the $OH\cdot$ radical, then a linear dependence of $[Fe^{3+}]$ on $[Fe^{2+}]$ would have been observed.

$$Fe^{2+} + \cdot OH \rightarrow Fe^{3+} + OH^- \tag{5.4}$$

The above reaction (i.e. oxidation of Fe^{2+}) is perhaps one of the most reliable and widely used of all available systems for chemical dosimetry. It is known as the Fricke [9] dosimeter and prior to the interest in sonochemistry it was used mainly in ionising radiation studies to determine the amount of absorbed energy. Recently Todd [10] has employed the technique to estimate G values (i.e. the number of radicals formed per 100 eV [$= 1.6 \times 10^{-17}$ J] of absorbed energy) for the H· and OH· radical. He estimates the values to be 0.1 and 0.35 respectively. His results also indicate that only a very small fraction of the total acoustic energy is used for the formation of free radicals and that the yields are independent of the frequency, gas content of the solution and the intensity distribution of the ultrasonic field.

Another important aspect of Miller's work was the observation that reproducible results could only be obtained if a single cell was employed which was clamped in a rigidly fixed position relative to the source and never moved throughout the series of experiments. This is an important point which must be considered when conducting

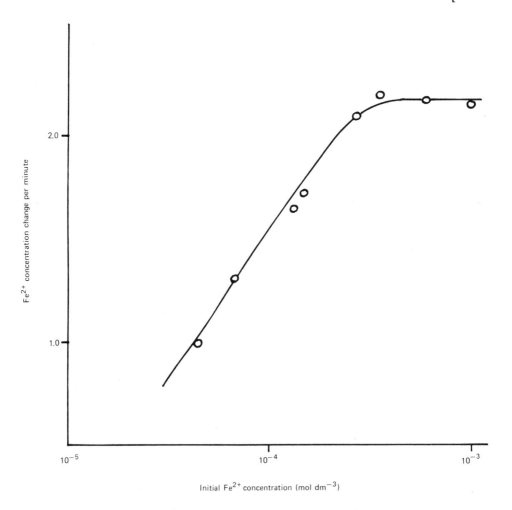

Fig. 5.1 — Rate of Fe^{2+} concentration change versus initial Fe^{2+} concentration.

ultrasonic experiments. Ultrasonic waves, being sound waves, are reflected at solid surfaces. Thus if an external source of ultrasound is employed the resultant available acoustic power will also be dependent on the geometry of the reaction vessel. According to Suslick, Schubert and Goodale [11], the use of a round-bottom flask in the cup-horn shows complex interference patterns in the sonochemical yield as a function of reaction vessel height (Fig. 5.2a). For flat-bottom vessels, regardless of the diameter, no interference pattern is observed (Fig. 5.2b). There still exists, however, as with the round-bottom vessel, an optimal solution height for maximum sonochemical effect. For the flat-bottomed vessel, the sonochemical yield is an order of magnitude more than that for the round-bottomed vessel.

Similar studies with the direct insertion probe indicate that neither the shape of the reaction vessel, nor the depth of immersion significantly affects the sonochemical

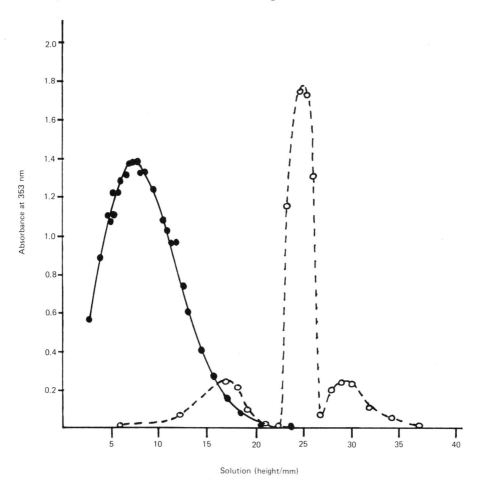

Fig. 5.2 — Ultrasonic irradiation of aqueous sodium iodide (1 mol dm^{-3}) absorbance after 2 minutes irradiation (22 kHz; 150 W; $T = 10°C$). (a) ○, Round-bottom flask (undiluted); (b) ●, flat-bottom flask (diluted tenfold).

yield. However, it must be remembered that for some reactions the extent of sonochemical mixing is determined by the shape of the reaction vessel — see Chapter 7, Rosett cell.

In an attempt to understand the mechanism by which intense ultrasonic waves caused chemical changes in liquid systems, Weissler [12] investigated the effect of various volatile scavengers on the yield of H_2O_2 in both oxygen- and argon-saturated aqueous solutions. Working at a frequency of 400 kHz and an intensity of 2.5 W cm^{-2} (determined calorimetrically), he observed that allylthiourea, formic acid and acrylamide all decreased the sonochemical yield of H_2O_2 (Fig. 5.3).

These results were in direct contrast to those obtained using radiation techniques. For example, Hart [13] had observed 30% and 70% *increased* yields in H_2O_2 formation in the presence of formic acid using Co^{60} γ-rays and 110 kV X-rays

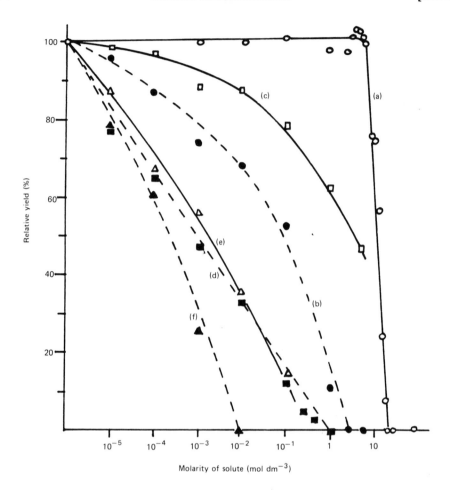

Fig. 5.3 — Relative yield of H_2O_2 as a function of radical scavenger concentration. (a) O, O_2/formic acid; (b) ●, Ar/formic acid; (c) □, O_2/acrylamide; (d) ■, Ar/acrylamide; (e) △, O_2/allylthiourea; (f) ▲, Ar/allylthiourea.

respectively. The increased yields of peroxide in the radiolysis studies were thought to be due to the suppression, by the scavengers, of reactions 5.5 and 5.6 (which effectively lower the H_2O_2 yield) and the increased likelihood of reaction 5.7 occurring.

$$H\cdot + H_2O_2 \rightarrow H_2O + \cdot OH \qquad (5.5)$$

$$H_2 + \cdot OH \rightarrow H_2O + H\cdot \qquad (5.6)$$

$$H\cdot + \cdot OH + HCOOH + O_2 \rightarrow H_2O_2 + H_2O + CO_2 \qquad (5.7)$$

The inference must be that different mechanisms are operating in the two systems.

A closer examination of Fig. 5.3, which compares the efficiency (observed H_2O_2 yield divided by the yield at zero concentration of scavenger) of the 3 scavengers as functions of scavenger concentration, also reveals two interesting features. The first observation is that the % efficiency of any radical is less in the presence of oxygen than in the presence of argon. This, according to Weissler, is a consequence of the increased possibility of the alternative route to peroxide formation via reactions (5.8) and (5.9), rather than the usual recombination route (eqn 5.2). Such a suggestion is confirmed by Jeager's work [14] with isotopic oxygen.

$$HO\cdot + \cdot OH \rightarrow H_2O_2 \tag{5.2}$$

$$H\cdot + O_2 \rightarrow HO_2\cdot \tag{5.8}$$

$$HO_2\cdot + HO_2\cdot \rightarrow H_2O_2 + O_2 \tag{5.9}$$

The second feature is the observation that formic acid proved to be a less efficient scavenger than either acrylamide or allylthiourea. This is unexpected since formic acid being more volatile, ought to find it easier to enter the vapour in the microbubble. Such an observation suggests that radical recombination (to yield H_2O_2) takes place partly in the liquid phase. Although this view is supported by Henglein (see later) it was in contrast to the findings of Anbar and Pecht [15] who found that whereas the non-volatile $HO\cdot$ scavengers, thallous and formate ions, hardly affected the yield of H_2O_2, a number of different volatile organic solutes (e.g. methanol) significantly lowered the yield of H_2O_2 — Table 5.1. The inference is that

Table 5.1 — The sonochemical formation of $H_2O_2^{18,18}$ in H_2O^{18} under argon in the presence of $H_2O_2^{16,16}$

Sonolysis (Time/min)	H_2O_2 $(10^{-3}\,mol\,dm^{-3})$	Additive	Additive $(10^{-2}\,mol\,dm^{-3})$	$[H_2O_2]^a$ $(10^{-6}\,mol\,dm^{-3}\,min^{-1})$
20	2.5	—	—	4.95
60	2.5	—	—	4.95
80	6.3	—	—	5.05
120	2.7	—	—	4.75
85	2.0	Tl_2SO_4	1.0	4.80
90	1.9	Tl_2SO_4	2.0	4.00
100	1.9	Tl_2SO_4	5.0	3.70
93	1.9	Tl_2SO_4	15.0	3.30
85	2.5	HCOONa	5.0	3.80
10	2.3	HCOONa	50.0	4.50
100	2.3	MeOH	10.0	0.58
100	2.3	MeOH	50.0	0.06
81	2.3	O_2	2.5%	3.60

^{a}Normalised to 100^{18}%.

if H_2O_2 is produced by the radical recombination of $HO\cdot$ radicals (eqn 5.2), then recombination does not take place in the liquid but in the bubble phase since only the volatile components can enter the bubble.

Parke and Taylor [16] have demonstrated the presence of HO· radical by irradiating air, oxygen, and nitrogen saturated solutions of benzoic acid at 500, 1000 and 2000 kHz ($I = 1.2$ to 4.1 W cm^{-2}) to produce *ortho*, *meta* and *para*($o > p > m$) hydroxybenzoic acid. The authors also reported that hydroxylation took place for toluene, nitrobenzene and phenol, though not for benzene itself. The presence of H· radicals has been demonstrated independently by both Henglein, and by Anbar and Pecht. Irradiating water in the presence of D_2 as a scavenger, Henglein [17] observed initially an increase in the amount of HD produced which reached a maximum and then decreased as the amount of D_2 increased (Fig. 5.4).

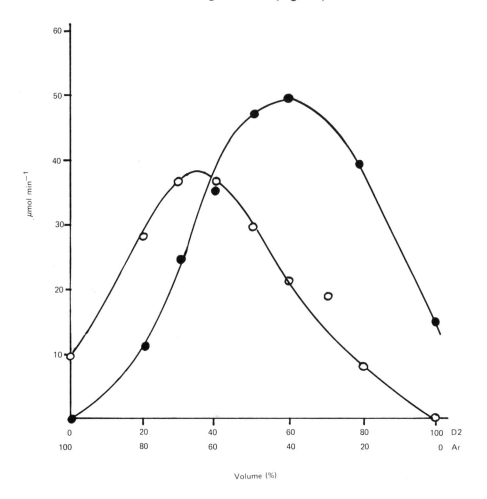

Fig. 5.4 — Rates of formation of H_2(O) and HD (●) under various mixtures of argon and deuterium.

Henglein explained the maxima in the yield versus gas composition curve (Fig. 5.4) in terms of two opposing effects. As the amount of D_2 is increased, the radical recombination reaction (eqn 5.10), which occurs very efficiently in the absence of a scavenger (e.g. D_2) is suppressed, and the primary radicals, H· and HO· are scavenged increasingly more efficiently to form HD and H_2 (eqns 5.11–5.14). However as the concentration of D_2 in the mixture increases beyond 50%, a lower temperature is reached in the adiabatic compression of the bubble and all yields decrease.

$$H· + ·OH \rightarrow H_2O \tag{5.10}$$

$$D_2 + ·OH \rightarrow HOD + D· \tag{5.11}$$

$$H· + D_2 \rightarrow HD + D· \tag{5.12}$$

$$H· + ·H \rightarrow H_2 \tag{5.13}$$

$$H· + ·D \rightarrow HD \tag{5.14}$$

To explain the similarity in shape for H_2, Henglein suggests that at low D_2 concentrations ·OH radicals are scavenged more efficiently than the H· radicals (eqn 5.11), thus allowing the H· radicals to provide larger yields of H_2 and HD (eqns 5.13 and 5.14). It must be emphasised, however, that Henglein presents no direct evidence for the formation of HOD.

Henglein [18, 19] has adopted the above ideas to explain the results obtained from other investigations into the effect of gas type and concentration on the sonolysis products of water, where again maxima were obtained in product yield versus gas composition. Perhaps the most significant contribution from Henglein's work on the effect of various gases, is the suggestion that at low concentration of the second gas (argon always present), there exists a similarity between the chemistry which takes place in the sonolytic reaction and that found to occur pyrolytically — i.e. in the sonolytic reaction the gas is acting as a reactant.

Henglein [20] has extended his earlier investigations on the effect of organic additives on the sonochemical yield of H_2O_2 and determined the concentration ($C_{1/2}$) of added solute needed to reduce the H_2O_2 yield by 50%. Although he was unable to obtain correlations between either the boiling point ($\propto$ -vapour pressure) and $C_{1/2}$, or $C_{1/2}$ and the rate constant for OH scavenging (as determined from radiation chemical studies in homogeneous solution), there did seem to exist a correlation between $C_{1/2}$ and an empirical measure of the hydrophobicity (R) of the scavenger (Fig. 5.5).

The factor R, according to Henglein, is the ratio of hydrophobic groups to hydrophilic groups of the solute molecules, i.e. the hydrophobic groups CH_3 and CH_2 were counted as 1.0, CH as 0.5, and *t*-C atoms as zero; the hydrophilic groups OH, COO–, CO, and NH_2 were counted as 1.0. According to our previous notions, hydroxyl radicals generated in the gas phase during transient cavitation could recombine (eqn 5.2) to form H_2O_2 in the gas phase, at the gas–liquid interface, or in the bulk of the solution. If most of the hydroxyl radicals recombined in the bulk of the solution, the scavenging efficiency should correlate with the rate constants for the reactions of the scavenger with hydroxyl radicals. The fact that there exists a correlation of $C_{1/2}$ with the hydrophobicity of the scavengers suggests that most of the radical recombinations occur in the gas phase or at least at the gas–liquid interface.

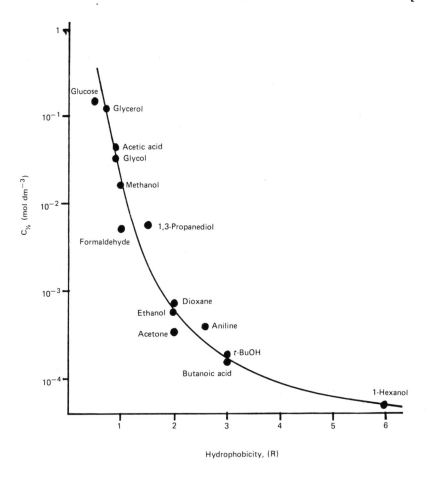

Fig. 5.5 — Concentration ($C_{1/2}$) versus hydrophobicity (R).

Perhaps the most conclusive direct evidence for the formation of H· and OH· radicals by the ultrasonic irradiation of water is provided by Riesz [21]. In a series of experiments in which he irradiated 3 cm^{-3} samples of water (bath, 50 kHz, temperature 23°C, Intensity 600 W m^{-2}) containing various spin-trap molecules and through which was continually bubbled argon gas (0.5 dm^3 min^{-1}), he was able to trap out and identify (by e.s.r) the radicals H· and OH·. Further experiments of the competing effect of various radical scavengers (e.g. formate, thiocyanate, benzoate, methanol, ethanol, 1-propanol, 2-methyl-2-propanol, acetone and 2-methyl-2-nitrosopropane [MNP]) and the effect of the gas type on the yield of the spin-trapped adducts seems to confirm the physical principles identified in Chapter 2. For example, the presence of acetone and MNP was found to reduce greatly the formation of the adducts suggesting that either these scavengers, by virtue of their higher vapour pressures, had entered the bubble and thereby increased the scavenging efficiency, or that their presence in the bubble had cushioned the collapse during cavitation and thus reduced radical formation. In respect of experiments in the presence of various gases, the work indicated that dissolved gases of low thermal conductivity (argon) produced greater sonochemical yields than those of higher conductivity — see Fig 5.6. The overall conclusion is that insonation of water produces both H· and OH· radicals and these are produced in the cavitation bubble.

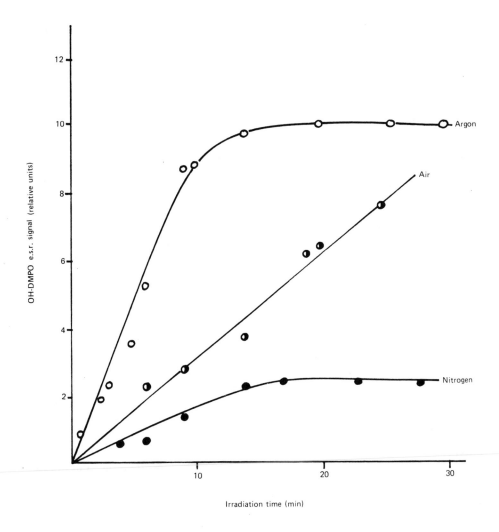

Fig 5.6 — Sonication time dependence of E.S.R. intensity of OH-DMPO adduct.

More recently spin-trap techniques have been used to detect the free radical intermediates formed during the decomposition of aromatic diazonium salts [22].

5.3 KINETIC ASPECTS

Of the few groups who have addressed themselves to detailed kinetic studies, most have considered reactions in aqueous rather than non-aqueous media. There is no doubt that until recently the lack of progress in organic media was the result of a combination of two factors (i) the failure to observe, in organic media, certain sonochemical reactions which occurred in water and (ii) the knowledge that the addition of organic solutes suppressed sonochemically-induced aqueous reactions. With hindsight it is probable that this earlier lack of success was due as a result of the higher vapour pressures of the organic liquids which, in turn, led to substantial lowering of the cavitational intensities in the organic media and therefore lack of reactivity. In recent times, with the advent of more powerful instrumentation (see Chapter 7) a resurgence of interest in non-aqueous studies has occurred, most notably in the field of synthetic organic chemistry, the subject of Chapter 3. In general non-aqueous high intensity ultrasound can be broadly divided into three major areas. These are cavitation-induced decomposition of the solute or solvent, ultrasonically-induced free radical polymerisation and ultrasonic polymer degradation. The latter two areas are dealt with in Chapter 4 (Polymers), and the first will be dealt with later (5.3.2).

5.3.1 Aqueous systems

Apart from our investigations into the effect of ultrasound on the solvolysis of 2-chloro-2-methylpropane, the most widely investigated aqueous system is the acid catalysed hydrolysis of methyl ethanoate. This system has been studied by several workers [2–4, 23], all of whom observed rate increases, albeit small (4–14%) on the application of ultrasound. The precise origin of these rate enhancements however have been attributed to different aspects of cavitation. For example, Folger and Barnes [4] (using 27.5 kHz and $I = 2.5$–4.4 W cm^{-2}) have attributed the increased reaction rate to the large temperature effects associated with the microbubbles, whereas Couppis and Klinzing [2] (at 540 kHz and 780 kHz) and separately Chen and Kalback [3] (at 23 kHz) have attributed the increases to increased molecular motion of the reactants. This latter conclusion is a consequence of the authors' observations that whereas the activation energies in the absence and presence of ultrasound were similar (Figs 5.7 and 5.8), the intercept values (A) for the Arrhenius plots were different. If, according to the Absolute Rate Theory, the reaction frequency factor (A) is related to vibratory motion of the reacting molecules, then the generation of the tremendous local pressure gradients during cavitation may increase the vibratory motion.

Support for the ideas of Folger and Barnes (i.e. that enhanced reactivity is due to the high bubble temperature) is demonstrated by the work of Griffing [24]. Griffing studied the effect of high frequency ultrasound (2 MHz, $I = 6.5$ W cm^{-2}) on the acid catalysed inversion of sucrose and observed no detectable rate enhancements, even though the solution visibly cavitated. The particular reaction was chosen since it proceeded at a reasonable rate at room temperature and had a high temperature coefficient — for this reaction a rise in temperature of 10°C brings about a threefold increase in rate. Since sucrose is a solid and therefore has a negligible vapour pressure, it cannot enter the cavitating bubble and hydrolysis must take place in the liquid phase. The lack of enhancement with ultrasound was interpreted by Griffing as

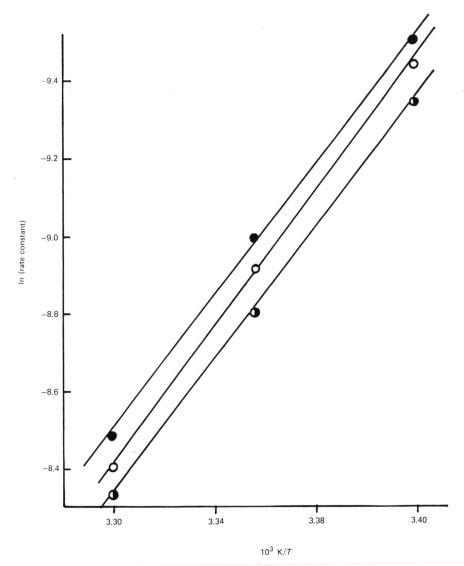

Fig. 5.7 — Arrhenius plots for methyl ethanoate hydrolysis at various transducer voltages
(Reference 2). ●, zero volts; ○, 80 volts; ○, 105 volts.

indicating that there was no appreciable heating at the bubble–liquid interface. That
increased reactivity is not due simply to an increase in the bulk temperature of the
system (due to ultrasonic heating) is afforded by the work of Kristol, Klotz and
Parker [1].

Kristol has reported ultrasonically (20 kHz) induced rate enhancements (~ 14%)
for the hydrolysis of the 4-nitrophenyl esters of a number of aliphatic carboxylic
acids. If bulk heating were responsible then considerable variation in the rate

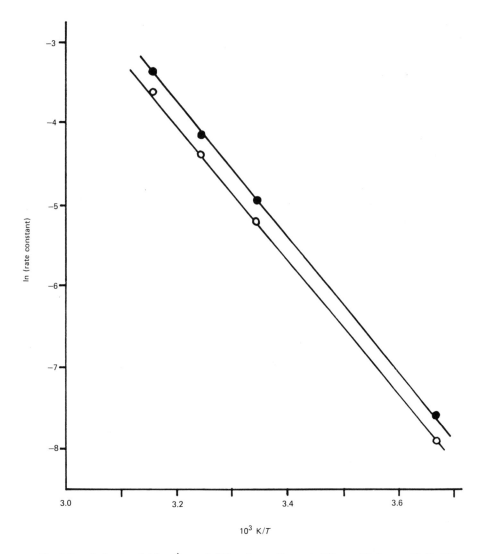

Fig. 5.8 — Arrhenius plot for ultrasonic (●) and non-ultrasonic (○) runs (Reference 3). ●, 450 mA; ○, zero mA.

enhancements would have been anticipated, not the constant 14% observed, due to the large differences in activation energy for the hydrolysis of each substrate. The rate enhancements were thought to be most probably due to the intense localised pressure increases as a result of bubble collapse. Perhaps such conclusions prompt the question 'Are the rate enhancements due to increased concentration effects, increased diffusion or micro pressure-jump.?'

Doulah [23] has used the data obtained by Couppis and Klinzing to suggest that the observed rate enhancements in the aqueous methyl ethanoate system are the result of increased diffusion within the reacting system. The suggestion is that for this particular system, which is a fast reaction controlled by diffusion of the reacting ions in the solvent [25], the shock waves produced on bubble collapse influence the diffusion coefficients of the reactants. Consideration of these ideas led Doulah to derive equation (5.15)

$$\Delta K = C(V^2 - V_0^2)^{1/2} \tag{5.15}$$

where ΔK is the increase in rate constant on sonication, V is the voltage applied to the transducer, V_0 is the applied voltage necessary to cause the onset of cavitation and C is a constant related to the radius of the sphere of diffusion and the kinematic viscosity.

Verification of equation (5.15) is given in Fig. 5.9, where in the absence of relevant information, Doulah has arbitrarily set V_0 equal to zero. Notwithstanding this, the plots give reasonably good agreement with theory. The scatter at 293 K is thought to be due to deviations at this temperature from the linear relationship, assumed by Doulah, between cavitation energy and the power input to the sound source.

Couppis and Klinzing have also investigated the effect of intensity, frequency and reaction volume on the hydrolysis of methyl ethanoate. With regard to intensity they observed, at several temperatures, an optimum power at which maximum rate enhancement occurred — Fig. 5.10.

This phenomenon has also been observed by Folger and Barnes for the same experimental system, and by ourselves for the solvolysis of 2-chloro-2-methyl-propane in various alcohol–water mixtures, and as such confirms the theoretical principles outlined in Chapter 2. For example, if it is argued that the rate enhancements observed in these systems are caused by the pressure gradients created on bubble collapse, it is necessary to examine the parameters which affect such a collapse. It has already been stated that the collapse time τ for a bubble is

$$\tau = 0.915 \, R_m (\rho/P_m)^{1/2} (1 + P_{vg}/P_m) \tag{2.27}$$

and is therefore dependent upon R_m, the maximum size to which the bubble will grow during the rarefaction cycle. This maximum size however is governed by the applied intensity (I) or more correctly the acoustic amplitude, P_A.

$$I = P_A^2/2\rho c \tag{2.13}$$

$$R_{max} = (4/3w_a) \, (P_A - P_h) \, (2/\rho P_A)^{1/2} [1 + (2/3P_h) \, (P_A - P_h)]^{1/3} \tag{2.38}$$

Thus with increased intensity, the bubble radius may grow to such a size that the collapse time exceeds the time during which the compression cycle acts (i.e. the maximum time available for collapse) and the bubble will not totally collapse. This will obviously lead to a reduction in the number of cavitational events, and hence shock wave pressure, and consequently reduced rate enhancements.

In order to study the effect of frequency, Couppis and Klinzing investigated the

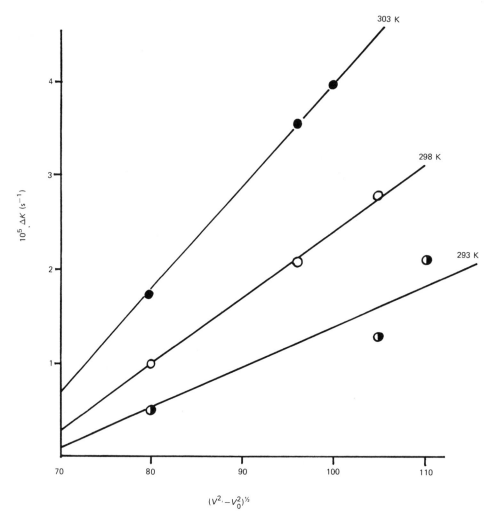

Fig. 5.9 — Increase in the hydrolysis rate constant (ΔK) versus cavitation energy expressed in terms of $(V^2 - V_0^2)^{1/2}$ for methyl ethanoate.

reaction at 540 and 780 kHz. As Fig. 5.11 shows, lower rate enhancements were obtained at the higher frequency, again confirming the theoretical predictions — i.e. increasing the frequency decreases the period of oscillation of the sound wave, thereby reducing the time available in which growth to maximum size can occur. Although Fig. 5.11 does not reveal an optimum power for maximum enhancement at 780 kHz, investigations with other different initial reaction volumes does — Fig 5.12. In fact the position of optimum power, especially at 540 kHz, is found to vary quite substantially with reaction volume — Fig. 5.13.

The most obvious conclusion is that there is an optimum power for each volume and temperature where the reaction rate is a maximum. At 298 K and 540 kHz the

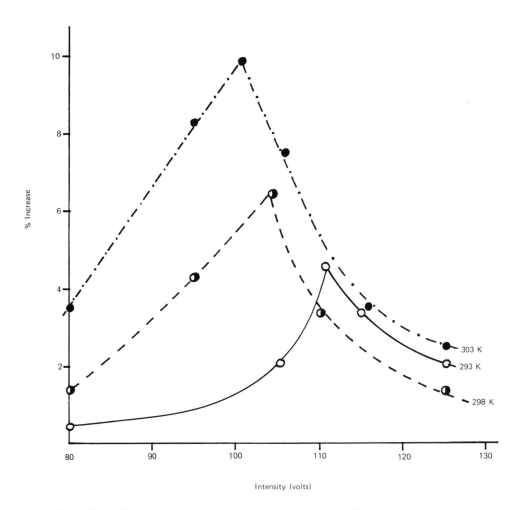

Fig. 5.10 — Effect of intensity and temperature on the conversion of methyl ethanoate.
Volume = 105 cm³; f = 540 kHz.

authors deduced an optimum power density, for all volumes, and obtained a constant value of $17 \times 10^4 \mathrm{W\ m^{-2}\ kg^{-1}}$.

Jozefowicz and Witekowa [26] have studied the effects of frequency, acoustic intensity, temperature and nature of the gas on a wide variety of aqueous redox reactions. Of the twelve reactions investigated eight were found to proceed, under irradiation, by zero order kinetics, an observation noted by ourselves for the aqueous polymerisation of N-vinyl pyrrolidinone [27]. Although the redox reaction rates were found to be independent of the frequency employed (15, 20, 25, 500, 800, 1000, 2100 kHz), maxima were observed when plotted as functions of temperature and intensity as predicted by theory (see General Principles). In the presence of various gases the authors observed the following sequence of rate constants:

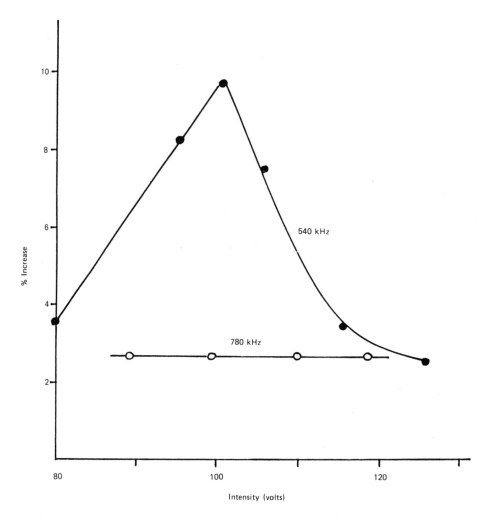

Fig. 5.11 — Effect of intensity and frequency on conversion of methyl ethanoate. $T = 303$ K; volume = 105 cm^3.

$k_{air} > k_{oxygen} > k_{argon} > k_{nitrogen}$. The larger rate constants in the presence of air and O_2 were explained in terms of the production of the oxidising radical, $HO_2\cdot$ (eqn 5.8). Several other investigations [28,29], though not so detailed, have also been reported.

One of the first kinetic studies of ultrasonically induced polymerisation was that reported by Lindstrom and Lamm in 1950 in which they investigated the polymerisation of acrylonitrile in dilute aqueous solution at several different acoustic frequencies and intensities. By assuming that the initiating species were either $H\cdot$ or $OH\cdot$ radicals (or both), produced during the cavitation process, or that radicals were produced as a result of polymer (P) degradation, the author proposed the following mechanism:

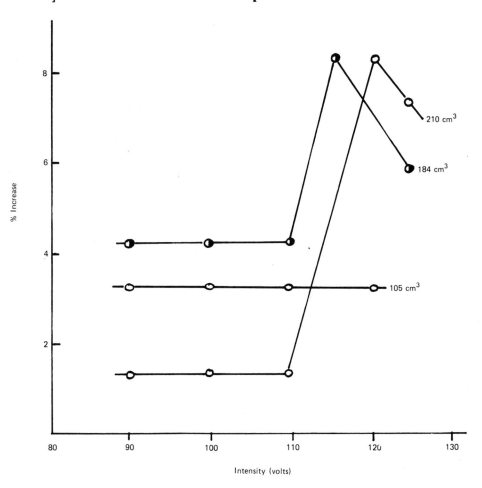

Fig. 5.12 — Effect of intensity and volume on the conversion of methyl ethanoate. $T = 303$ K;
$f = 780$ kHz.

$$H_2O \xrightarrow{k_1} H\cdot + \cdot OH \ (= 2R\cdot) \tag{5.16}$$

$$P \xrightarrow{k_2} 2R\cdot \tag{5.17}$$

$$R\cdot + nM \xrightarrow{k_3} R_n\cdot \tag{5.18}$$

$$R_n\cdot + R_m\cdot \xrightarrow{k_4} P \ (\text{dead polymer}) \tag{5.19}$$

On applying the steady state approximation, they deduced that the radical concentration, $[R\cdot]$, and propagation rate, R_p, were given respectively by:

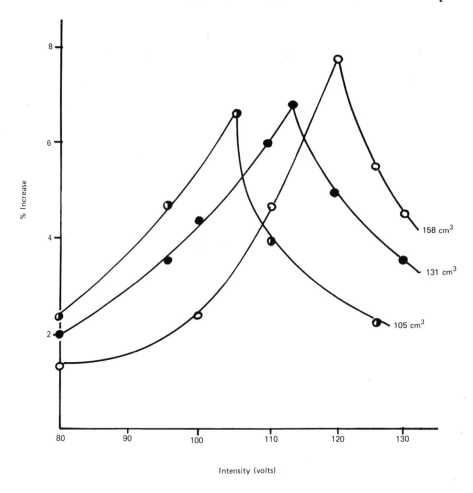

Fig. 5.13 — Effect of intensity and volume on the conversion of methyl ethanoate. $T = 298$ K; $f = 540$ kHz.

$$[R\cdot] = \left\{ \frac{k_1 + k_2[P]}{k_4} \right\}^{1/2} \qquad\qquad (5.20)$$

$$R_p = -d[M]/dt = k_3 [M][R\cdot] \qquad\qquad (5.21)$$

On expressing both the monomer and polymer concentrations in terms of p ($[M]_0 - [M])/[M]_0$, the fraction polymerised (or extent of reaction), equation (5.21) reduces to

$$\frac{dp}{dt} = (1-p)k_3 \sqrt{\frac{k_1 + [M]_0 k_2 p/n}{k_4}} \qquad\qquad (5.22)$$

where n is the average number of monomers in the polymer (i.e. $[P] = ([M]_0 - [M])/n$. For this particular polymerisation reaction, the extent of reaction over the time interval investigated never exceeded 0.02. By making the assumption that $(1 - p)$ could be approximated to 1, integration of equation (5.22) yielded equation (5.23), a relationship which the authors found fitted the data extremely well.

$$t = c_1 \{\sqrt{(1 + c_2 p)} - 1\} \tag{5.23}$$

where $c_1 = 2nk_1^{1/2} k_4^{1/2}/(k_2 k_3 [M]_0)$ and $c_2 = k_2 [M]_0/k_1 n$.

Kruus has suggested an alternative mechanistic and kinetic approach for polymerisation in non-aqueous media and this will be discussed later.

5.3.2 Non-aqueous systems

Although the first example of sonolysis in a non-aqueous solvent, the decolourisation of diphenylpicrylhydracyl (DPPH) radical in methanol, was reported in 1953, it took some twenty years to realise that cavitation could successfully be supported in organic solvents. The reasons for this have already been identified.

Perhaps one of the most significant contributions to the advancement of the understanding of the origins of sonochemical reactions in non-aqueous media is that provided by the work of Suslick et al. In their search to identify a more appropriate dosimeter system for non-aqueous media (in non-aqueous solutions iodide oxidation is no longer useful), Suslick et al. [28] investigated two alternative systems;(i) the bleaching of the radical scavenger diphenylpicryhydracyl, DPPH and (ii) the decomposition of iron pentacarbonyl, $Fe(CO)_5$. The former reaction was chosen since it is known that DPPH is a good radical scavenger which would scavenge any radicals produced in the sonolysis; the latter was chosen for investigation since both its thermal and photochemical reactivity were well known. In both cases Suslick et al. observed a reasonable correlation between ln(rate) and the solvent vapour pressure — Figs 5.14 and 5.15, Table 5.2. In fact this correlation can be predicted if it is assumed that the chemical reactions take place in the bubble.

According to the theory outlined in General Principles the maximum temperature reached inside a collapsing transient bubble may be taken to be inversely proportional to the solvent vapour pressure (P_v). If it can be assumed that the reaction mechanism in the bubble is governed by the Arrhenius equation (eqn 5.24), then it follows (from eqn 2.35) that the reaction rate constant (k) decreases with increasing solvent vapour pressure, P_v (eqn 5.25).

$$\ln k = \ln A - \frac{E_a}{RT} \tag{5.24}$$

$$\ln k = \ln A - \frac{E_a . P_v}{RT_o P_m(\gamma - 1)} \tag{5.25}$$

In other words the slopes of the lines give a measure of $E_a/RT_o P_m(\gamma - 1)$, where T_o is the ambient (experimental) temperature, P_m is the pressure in the liquid at the

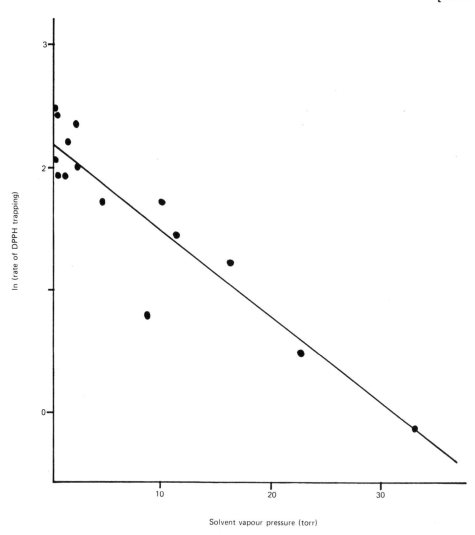

Fig. 5.14 — Plot of ln (rate of DPPH trapping) versus solvent vapour pressure. (Data in Table 5.2).

moment of transient collapse ($= P_h + P_a$) and the other symbols have their usual meaning. Suslick, Hammerton and Kline [29] have extended their investigations in an attempt to determine the actual site of the sonochemical reaction — the so-called 'hot-spot'. Using as solvent a mixture of two n-alkanes, mixed in such a way as to keep an *overall* constant system vapour pressure (5 torr) at each experimental temperature, and employing a *fixed* concentration of the metal carbonyl (0.01 M), they have determined the first order rate constant for ligand exchange with triphenyl-phosphine as a function of the vapour pressure of the metal carbonyl, P_v. On plotting the rate constant (k_{obs}) versus P_v, Suslick, Hammerton and Kline obtained a

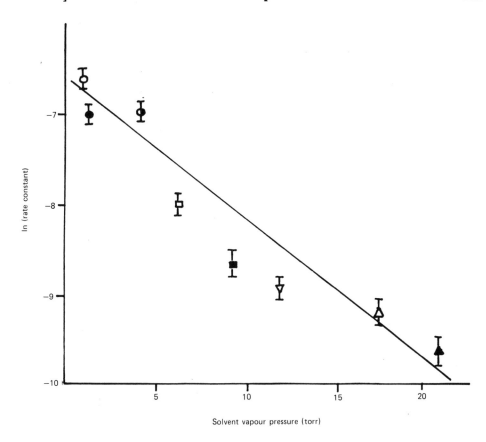

Fig. 5.15 — Ln (rate of $Fe(CO)_5$ decomposition) versus solvent vapour pressure. $T = 298$ K; intensity $= 80$ W cm^{-2}; argon gas. $\bigcirc$, Decalin; $\bullet$, Decane; $\bigcirc$, Nonane; $\square$, 0.22 mol fraction octane in nonane; $\blacksquare$, 0.52 mol fraction octane in nonane; $\triangledown$, Octane; $\triangle$, 0.11 mol fraction heptane in octane; $\blacktriangle$, 0.22 mol fraction heptane in octane.

non-zero intercept — Fig. 5.16. The assumption was that as the substrate vapour pressure (P_v) increased, the concentration of the substrate in the bubble increased thus increasing k_{obs}. The non-zero intercept (i.e. vapour pressure independent component of the rate) was taken as a measure of the rate in the liquid phase. From such assumptions Suslick, Hammerton and Kline have been able to deduce that the temperature in the bubble and at the interface are 5200 K and 1900 K respectively.

Donaldson, Farrington and Kruus [30] have also obtained a similar, though somewhat less successful correlation of ln(rate) versus P_v for the polymerisation of nitrobenzene, at 20 kHz and 20 W cm^{-2}, in the presence of various gaseous and liquid solutes. The reaction rate was not only lowered by the presence of a liquid solute with high vapour pressure, it was also lowered by the presence of a gaseous solute with a high solubility. Experiments by Kruus and Patraboy [31] with methyl methacrylate (MMA) have shown that intense ultrasound (20 W cm^{-2}) can be used to initiate the polymerisation of a pure vinyl monomer. Although the yields obtained

Table 5.2 — Rate of DPPH trapping in non-aqueous solvents

Solvent	Vapour Pressure (torr)	$-\mathrm{d[DPPH]}/\mathrm{d}t$ (μmol dm^{-3} min^{-1})
1-Hexanol	0.12	7.74
Decalin	0.20	11.8
Decane	0.25	7.75
1-Pentanol	0.35	6.66
1,3,5-Trimethylbenzene	0.53	11.4
1-Butanol	1.1	6.20
Cyclohexanone	1.3	8.91
Di-n-butylether	1.9	10.4
1,3-Dimethylbenzene	2.2	7.25
1-Propanol	4.6	5.43
Toluene	8.7	2.17
2-Hexanone	10.1	5.44
2-Pentanone	11.3	4.13
Isooctane	16.4	3.37
Di-n-propylether	22.8	1.61
2-Butanone	33.1	0.88

Rates determined after 10 minutes irradiation at 50 W cm^{-2} under Argon at 4°C; initial [DPPH] = 2×10^{-4} mol dm^{-3}.

are too small ($<4\%$/h) to suggest that the technique can provide a commercial alternative to the conventional methods, the investigation is important in that it is one of the rare studies into the effects of acoustic intensity on the rates of ultrasonically induced polymerisation. It also dispels the view, held by previous workers, that a polymerisation reaction could only occur in the presence of polymer. To explain the observed variations of polymerisation rate with time (Fig. 5.17), reaction volume (Fig. 5.18) and acoustic intensity (Fig. 5.19), Kruus and Patraboy adopted a mechanism similar to that of Lindstrom and Lamm with the exception that the cavitation bubbles were regarded as a reactant.

$$M + C \xrightarrow{k_1} 2R_0 \cdot \quad (slow, RDS, initiation) \qquad (5.26)$$

$$R_0 \cdot + M \xrightarrow{k_2} R_1 \cdot \quad (fast) \qquad (5.27)$$

$$R_n \cdot + \overset{\cdot}{M} \xrightarrow{k_3} R_{n+1} \cdot \quad (propagation) \qquad (5.28)$$

$$R_n \cdot + R_m \cdot \xrightarrow{k_4} P \quad (dead\ polymer,\ termination) \qquad (5.29)$$

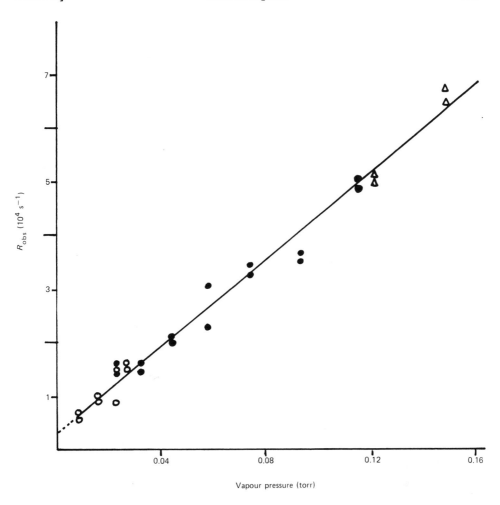

Fig. 5.16 — Observed first order rate constants for the sonolysis of $Fe(CO)_5$ versus vapour pressure of $Fe(CO)_5$. $\bigcirc$, Nonane/dodecane mixtures, 39°C to 43°C; $\bullet$, Octane/decane mixtures, 12°C to 39°C; $\triangle$, Heptane/nonane mixtures, -3°C to 14°C.

$$P + C \xrightarrow{\ k_5\ } 2R_n \cdot \quad \text{(slow, reinitiation)} \tag{5.30}$$

By applying the steady state analysis and assuming the concentration of cavitation bubbles, [C], could be expressed as:

$$[C] = \frac{N_c}{V} = \frac{FI}{V} \tag{5.31}$$

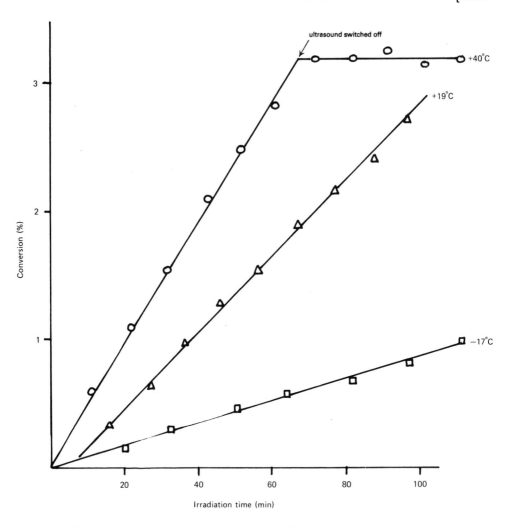

Fig. 5.17 — Monomer conversion as a function of irradiation time ($I = 20.6$ Wcm^{-2}).

where N_c is the number of cavitation bubbles, V is the reaction volume, I is the acoustic intensity and F is a proportionality constant, Kruus and Patraboy were able to deduce that the fractional conversion, $x(t)$, could be given by:

$$x(t) = \frac{(fk_1)^{1/2}k_3}{(2k_4)^{1/2}} \left(\frac{2FI}{V}\right)^{1/2} [M]_0^{1/2}t + \frac{k_3^2\, k_5}{8k_4}\left(\frac{2FI}{V}\right)\left(\frac{[M]_0}{\bar{X}_n}\right) t^2 \tag{5.32}$$

Although equation (5.32) predicts a quadratic time dependence for the conversion, the extremely good linear dependence exhibited by the data (Fig. 5.17) suggests

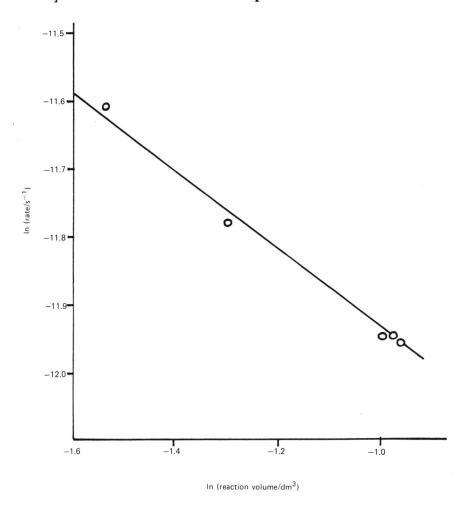

Fig. 5.18 — Ln (rate) versus ln (reaction volume).

that the second term of the equation is negligible, and that at these low conversions very few initiating radicals are produced via degradation of the polymer. If so, then the stationary concentration of radicals and the propagation rate (R_p) can be represented by:

$$[R\cdot] = \left(\frac{2fk[M][C]}{k_4}\right)^{1/2} \tag{5.33}$$

$$R_p = K[M]^{3/2}\left(\frac{FI}{V}\right)^{1/2} \tag{5.34}$$

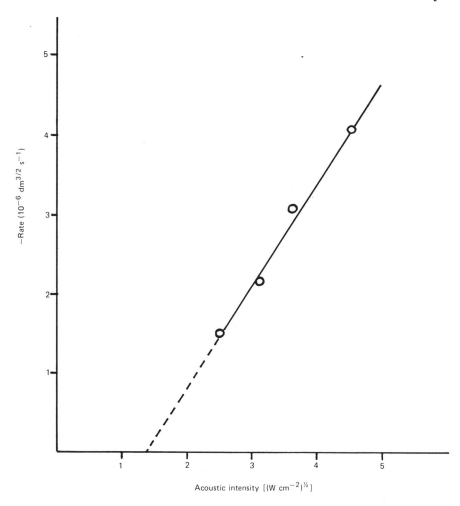

Fig. 5.19 — Polymerisation rate as a function of the square root of acoustic intensity.

where K is a constant $[= (2fk_1/k_4)^{1/2}k_3]$, and f is an efficiency factor introduced to represent the fraction of radicals which actually initiate the polymerisation.

Equation (5.34) predicts that the rate of polymerisation will be inversely proportional to the square root of the reaction volume, provided all other parameters are kept constant, a relationship verified by Fig. 5.18. When this volume dependence was used to correct the observed initial R_p values at the various initial reaction volumes, constant R_p values were obtained (Table 5.3). To test the validity of equation (5.34) for intensity, Kruus plotted the corrected polymerisation rates (compensated for the reduction in volume on removing aliquots from the sonicated reaction) against the acoustic intensity — Fig. 5.19. Several polymerisations were also carried out at constant intensity (20 W cm^{-2}) in the temperature range -17°C to 40°C. The Arrhenius plot yielded an activation energy of 19.3 kJ mol^{-1}, a value

Table 5.3 — Corrected and uncorrected polymerisation rates for various initial monomer volumes

Monomer (Volume dm^3)	Polymerisation rate	
	Uncorrected (10^{-6} s^{-1})	Corrected (10^{-6} dm$^{-3/2}$ s^{-1})
0.214	9.07 ± 0.08	4.14 ± 0.07
0.275	7.58 ± 0.10	4.03 ± 0.07
0.385	6.45 ± 0.08	4.08 ± 0.09
0.387	6.52 ± 0.10	4.05 ± 0.09
0.395	6.45 ± 0.10	4.08 ± 0.08

which was in good agreement with that reported (17–20 kJ mol^{-1}) for the bulk polymerisation of MMA when the contribution from initiation was excluded — i.e. initiation by ultrasound is independent of temperature, has negligible activation energy and in this respect resembles photopolymerisation.

Kruus also conducted experiments in the presence of the radical scavenger DPPH and observed induction periods (Fig. 5.20) which were roughly proportional to the concentration of DPPH employed. This clearly demonstrates the free radical nature of the polymerisation. By assuming that each of the monomer radicals produced by the cavitation process (eqn 5.26) reacted with one DPPH molecule he was able to deduce the following kinetic relationship

$$-\frac{d[\text{DPPH}]}{dt} = 2fk_1[M]_0 \left(\frac{FI}{V}\right) \tag{5.35}$$

which on integration yields

$$[\text{DPPH}]_0 - [\text{DPPH}]_t = 2fk_1[M]_0 \left(\frac{FI}{V}\right) t \tag{5.36}$$

Excellent correlations were obtained (>0.97) for plots of the consumed DPPH with time, intensity and reciprocal volume.

Analogous experiments were also performed at 21 W cm^{-2} in the temperature range $-17°C$ to $40°C$, but in the presence of DPPH. Although the data fitted the Arrhenius relationship it gave an effective activation energy of -85 kJ mol^{-1}. This observation prompted Kruus to fit the data to the modified Arrhenius (eqn 5.25) suggested by Suslick, where again an excellent correlation was observed between ln(rate) and P_v/T. (Fig. 5.21).

If it can be assumed that the cavitation bubbles are filled mainly with argon gas ($\gamma = 1.67$), the experiments were conducted with a gas flow rate of 20 cm^3 s^{-1}, and

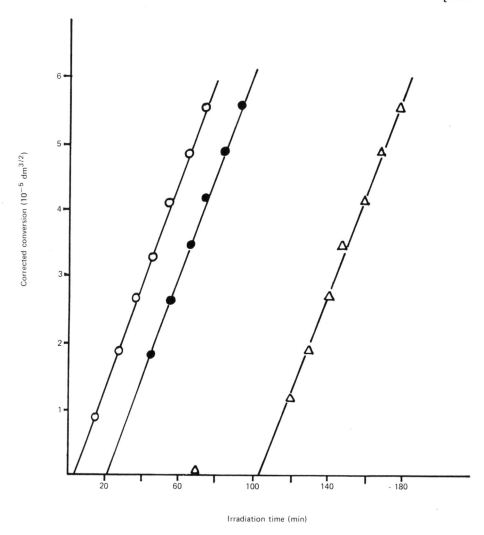

Fig. 5.20 — Effect of DPPH concentration on the corrected conversion–time curves. $T = 32°C$; intensity $= 20.6$ W cm^{-2}; $f = 20$ kHz. $\bigcirc$, [DPPH] $= 0.0$ mol dm^{-3}; $\bullet$, [DPPH] $= 2.1 \times 10^{-6}$ mol dm^{-3}; $\triangle$, [DPPH] $= 5.2 \times 10^{-5}$ mol dm^{-3}.

that the pressure on bubble collapse, P_m, ($= P_A + P_h$) is approximately 8 atmospheres ($P_A^2 = 2I\rho c$ — eqn (2.13), the slope of line ($= E_a/RP_m(\gamma - 1)$) in Fig. 5.21 ($= 14 \, K$ torr^{-1}) provides an estimate for E_a of 460 kJ mol^{-1}, a value close to that for the bond energy of a C–C bond (345 kJ mol^{-1}). Obviously the presence of MMA ($\gamma = 1.05$ [32]) will lower the the value of γ for the 'gas' resident in the bubble thereby lowering the effective E_a. The generality of the analysis, even allowing for the assumptions, would indicate that the site of radical formation is indeed the cavitation bubble itself.

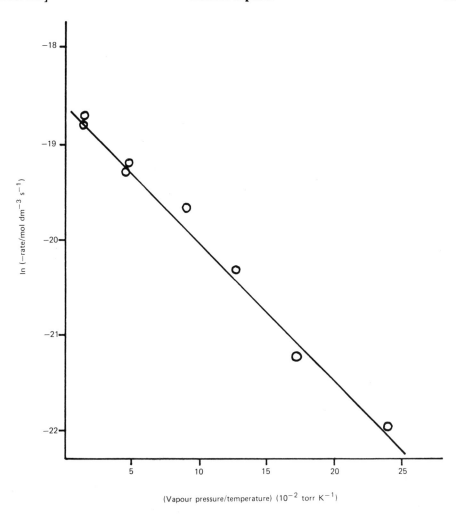

Fig. 5.21 — Plot of ln (DPPH rate) as a function of (vapour pressure/bulk temperature).

5.3.3 Mixed aqueous systems
The solvolysis of 2-chloro-2-methylpropane in a variety of mixed aqueous media has been the subject of numerous kinetic [33–35] and thermodynamic [36, 37] studies since it is a classic example of a unimolecular reaction (S_N1). Using a cuphorn reactor, operating at 20 kHz and delivering an acoustic intensity of approximately 1 W cm^{-2}, we have observed that ultrasonic irradiation has been found to accelerate the solvolysis, the effects being more marked at the lower temperatures and the higher alcohol compositions — e.g. a 20-fold rate acceleration occurring at 10°C and 60% w/w. This point is more clearly demonstrated in Fig. 5.22 where the rate enhancements (k_{ult}/k_{non}), for several alcohol–water mixtures, have been plotted against temperature.

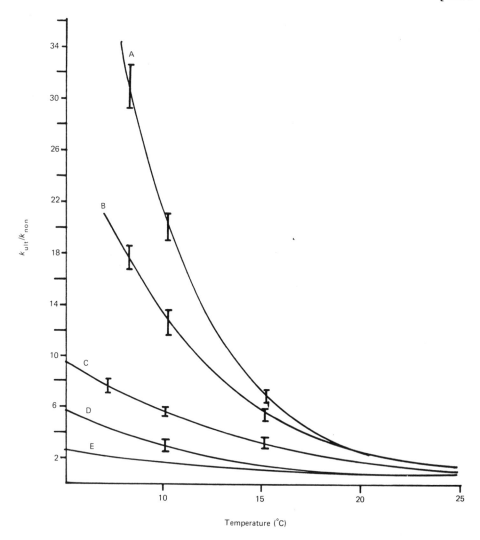

Fig. 5.22 — Ultrasonic rate enhancement versus temperature for different ethanol–water compositions. A, 60% EtOH–H$_2$O; B, 50% EtOH–H$_2$O; C, 40% EtOH–H$_2$O; D, 30% EtOH–H$_2$O; E, 20% EtOH–H$_2$O.

When the rate enhancements, at the various temperatures, are plotted against solvent composition (Fig. 5.23), a maximum in the curve occurs at 20°C. The composition at which this maximum occurs coincides reasonably closely with the maxima found in the viscosity (η) — composition and enthalpy of mixing (ΔH^m) — composition curves (Fig. 5.24). The existence of the exothermic maximum in ΔH^m is thought to be due to an increase in the hydrogen bonding within the solvent system. In simple terms we can envisage water as a 3D network, full of holes, in equilibrium with free water. The introduction of small amounts of ethanol, to water, is accompanied by the filling of these holes to produce a stronger structure and an

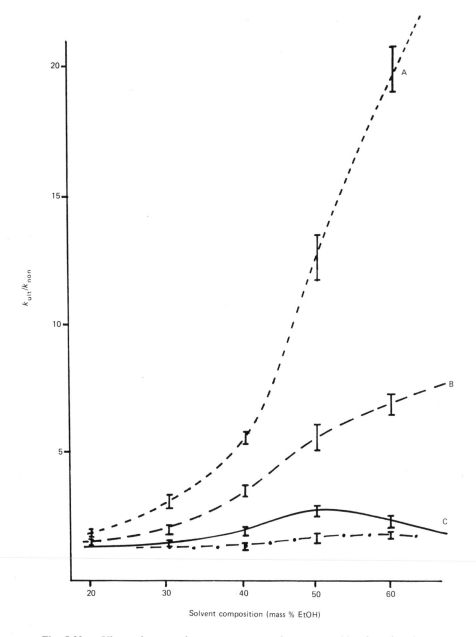

Fig. 5.23 — Ultrasonic rate enhancement versus solvent composition for ethanol–water at different temperatures. A, 10°C; B, 15°C; C, 20°C; D, 25°C.

increase in hydrogen bonding. Obviously at some composition, here 35% at 298 K, the alcohol can no longer be accommodated by the water and the structure begins to break down producing more 'free' water. At lower temperatures this point occurs at

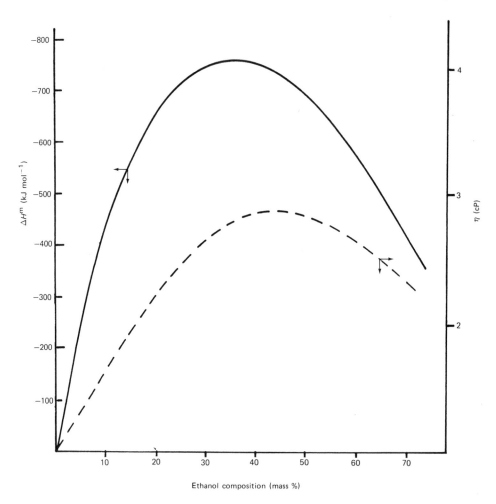

Fig. 5.24 — Relationship between enthalpy of mixing (ΔH^m), viscosity (η) and solvent composition.

a higher alcohol content. Perhaps of more significance is the existence of a maximum in the sound absorption composition curve. Figure 5.25 shows both the variation in sound absorption [38] and rate enhancement with solvent composition for ethanol--water mixtures. The tentative suggestion which has been offered for the existence of this maximum in the sound absorption curve is that the passage of the sound wave through the medium causes the translational energy to couple with the lowest vibrational mode to give a vibrationally excited state. Essentially then it is breaking the cooperative structure. The greater the structure, the more there is to break, and the greater the absorption. Hence if water is an equilibrium mixture of structured and free water, the application of ultrasound to the medium will drive the equilibrium in favour of 'free' water. The medium is apparently 'more aqueous' and will

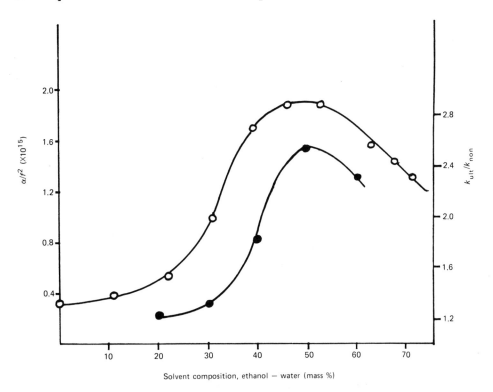

Fig. 5.25 — Variation of excess sound absorption (○) and rate enhancement (●) with solvent composition at 293 K.

provide an increase in the solvolysis rate. As a qualitative picture it seems quite reasonable. The simple deduction is that we should identify that solvent composition, for a binary aqueous mixture, where there is a maximum in the sound absorption curve and perform sonochemistry at that composition, since this ought to yield the greatest rate enhancement.

At first glance the rate, and absorption data for the t-butanol system (Fig. 5.26) seem to confirm this suggestion. Maxima exist for both the rate enhancement and sound absorption composition curves at approximately 30–35% (w/w) t-butanol. While it is gratifying to find that this relationship holds, closer inspection reveals an anomaly. The maximum in sound absorption in the t-butanol/water system (Fig. 5.26) is some 20 times greater than that in the ethanol/water system leading to the expectation of a correspondingly larger ultrasonic rate enhancement. Unfortunately the magnitude of the ultrasonic rate enhancement at this point is in fact 10-fold less.

The fact that ultrasound produces cavitation bubbles, which ultimately can be assumed to be filled with solvent vapour, led us to analyse the kinetic data in terms of solvent vapour pressure (eqn 5.25). If sonication simply provides 'hotspots' which accelerate the rate, then k (eqn 5.25) may be replaced by $(k_{ult}-k_{non})$. For the ethanol–water system, this is represented graphically by Fig. 5.27.

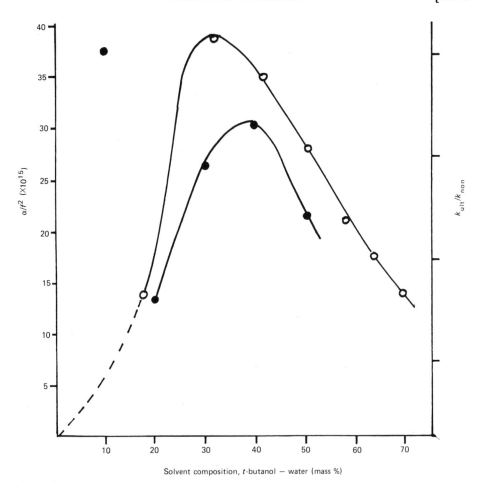

Fig. 5.26 — Variation of excess sound absorption (○) and rate enhancement (●) with solvent composition at 298 K.

Two points are evident:

(1) Consideration of the isothermal (variable composition) curves (...) reveals approximate linear relationships between $\ln(k_{ult}-k_{non})$ and P_v. At each temperature the rate constants increase with decrease in P_v, as required by equation (5.25). It must be recognised, however, that both E and A are known to vary with solvent composition [33, 34].

(2) If equation (5.25) is generally applicable, it would be expected that the isocomposition (variable temperature) curves (——) would show a better linearity, in that both A and E are constant for constant composition. However, even allowing for errors inherent in determining the rate constants it is difficult to concede that straight lines can be drawn for 20%, 30%, and 40% ethanol–water. The best approximations to straight lines are those to be found for 50% and 60%

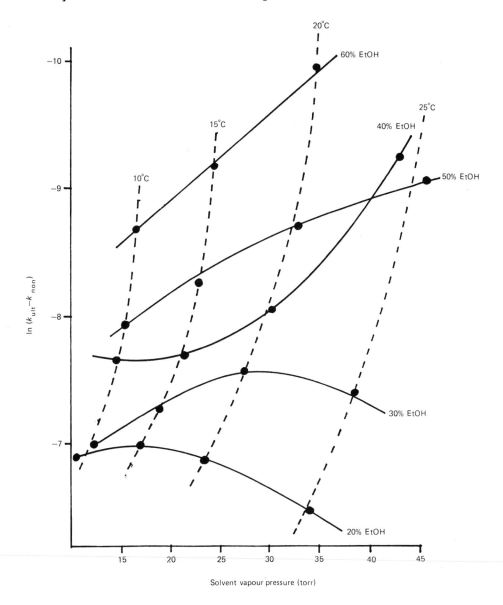

Fig. 5.27 — Variation in ultrasonic enhancement with solvent vapour pressure for the solvolysis of 2-chloro-2-methylpropane in ethanol–water mixtures.

(w/w) ethanol–water mixtures, compositions at which structural breakdown is thought to be occurring [39].

Consideration of a similar plot for *t*-butanol (Fig. 5.28) is somewhat better. Again there is approximate linearity for the isothermal curves with linearity for the isocomposition (constant temperature) curves occurring at approximately 20%

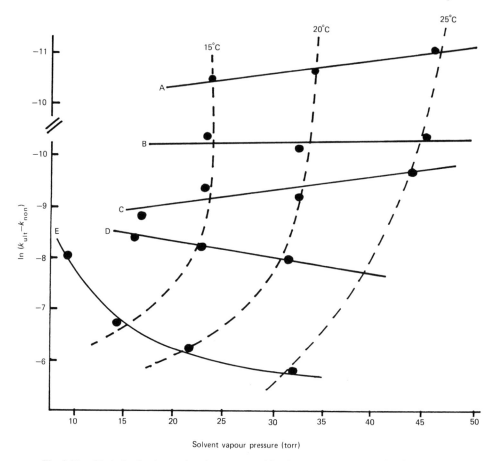

Fig. 5.28 — Variation in ultrasonic enhancement with solvent vapour pressure for the solvolysis of 2-chloro-2-methylpropane in *t*-butanol–water mixtures. A, 50% *t*-BUOH; B, 40% *t*-BUOH; C, 30% *t*-BUOH; D, 20% *t*-BUOH; E, 10% *t*-BUOH.

(w/w) butanol–water. Independent evidence suggests structural breakdown of *t*-butanol–water mixtures occurs beyond 25% (w/w) *t*-butanol. It may be fortuitous that the onset of linearity, for both systems, occurs beyond the region of maximum structure; on the other hand it may indicate that equation (5.25) is only applicable to non- structured (i.e. non-hydrogen bonded) solvents or solvent mixtures.

Having failed to obtain linear relationships between $\ln(k_{ult}-k_{non})$ and P_v, it was decided to assume that ultrasound was acting in a similar fashion to catalysis by lowering the activation energy for the reaction. This meant that no account had to be taken of the thermal contribution, since when comparing a catalysed system (here the reaction in the presence of ultrasound) with its non-catalysed system (here the conventional solvolysis), it is simply a question of comparing the separate Arrhenius plots and noting by how much the activation energy barrier is lowered. Figure 5.29

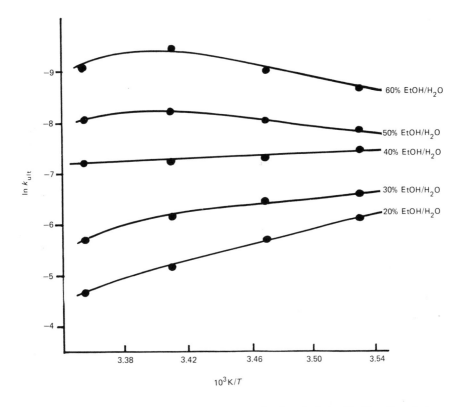

Fig. 5.29 — Ultrasonic rate constant versus temperature for the solvolysis of 2-chloro-2-methyl propane in ethanol–water mixtures.

shows the Arrhenius plots for the solvolysis reaction in various ethanol–water mixtures in the presence of ultrasound. Apart from 40% (w/w) ethanol–water all the curves deviate from linearity at the highest experimental temperature. Notwithstanding this, the other three points fit a good straight line and allow calculation of the activation energy and pre-exponential factor in the presence of ultrasound. These values together with those in the absence of ultrasound, are given in Table 5.4.

There are two points of interest.

(1) The values of E_{ult} decrease steadily as the composition is varied. Unlike the non-irradiated system there is no minimum; the values decrease monotonically. At 50% and 60% (w/w) the values of E_{ult} are negative, suggesting solvolyses in these mixtures should be instant. However, if it is assumed $E_{\text{ult}} = \Delta H^{\ddagger}_{\text{ult}}$ (the enthalpy of activation in the presence of ultrasound), then the negative $\Delta H^{\ddagger}_{\text{ult}}$ is more than compensated by the large negative $\Delta S^{\ddagger}_{\text{ult}}$ accompanying the reaction.

(2) Extrapolation of E_{ult} to 0% (w/w) provides a theoretical value, in water, of 110 kJ mol^{-1}. This value compares favourably with the conventional value of 100 kJ mol^{-1} found by independent workers — Winstein, and Robertson. The value of

Table 5.4 — Activation energies (kJ mol^{-1} and entropy values (J mol^{-1} K^{-1}) for ethanol–water mixtures

Ethanol (mass%)	E_{ult}	E_{non}	S_{ult}	S_{non}
20	62	87.9	− 75	7.5
30	30	81.9	− 193	10.3
40	10	85.2	− 270	− 18.9
50	− 21.6	89.4	− 386	− 19.6
60	− 49.5	96.2	− 491	− 2.03

A_{ult} however (obtained from the intercept), is very different from that obtained conventionally, and confirms the findings of Couppis and Klinzing, Chen and Kalback, and Folger, who in studying the effect of ultrasound on the acid hydrolysis of methyl acetate found very little variation in the energy of activation for the irradiated and non-irradiated reaction, yet substantially different intercept values (i.e. A_{ult} and A_{non}) from their Arrhenius plots.

Failure to fit the rate data to equation (5.25) (i.e. the ln$k$ versusP_v relationship), coupled with the ability to provide different A values in the presence and absence of ultrasound might be taken as evidence that with the present systems the power levels are insufficiently high ($\sim$1 W cm^{-2}) to provide transient cavitation, but do produce stable cavitation. In that a maximum in rate enhancement was observed at an ethanol composition similar to that at which maxima in ΔH^m, η and sound absorption were observed, prompted us to consider that ultrasound simply perturbed the bonding within the reacting system, and that the kinetics were a direct result of this perturbation.

During the solvolysis of a neutral substrate, which proceeds via a highly polar transition state, as is the case here, a considerable degree of solvent reorganisation will occur in the activation process as a result of the development of electric charge. The extent of this reorganisation is thought best reflected in the heat capacity of activation $\Delta C_p^\ddagger$ [40]. If it is assumed that solvent–solvent interactions are the major contributors to heat capacity then any decrease in these interactions will lead to a reduction in the heat capacity values. 2-chloro-2-methylpropane is expected to produce weak solute–solvent interactions, but will enhance the structure of the solvent in its immediate vicinity, by increasing the solvent–solvent interactions. The charged transition state on the other hand will promote solute–solvent interactions, at the expense of the solvent–solvent interactions, to produce negative $\Delta C_p^\ddagger$ values, which should become increasingly more negative the more structured the solvent. If ultrasound is capable of destroying the weak ground state solute–solvent interactions and a proportion of the solvent–solvent interactions, but is not capable of destroying the stronger solute-solvent interactions in the charged transition state, then $\Delta C_p^\ddagger$ would be expected to be positive, becoming increasingly more so the more structured solvent. In other words the $\Delta C_p^\ddagger$ — composition curve in the presence of ultrasound should be a mirror image of the curve in the absence of ultrasound. This view seems to be supported by Fig. 5.30.

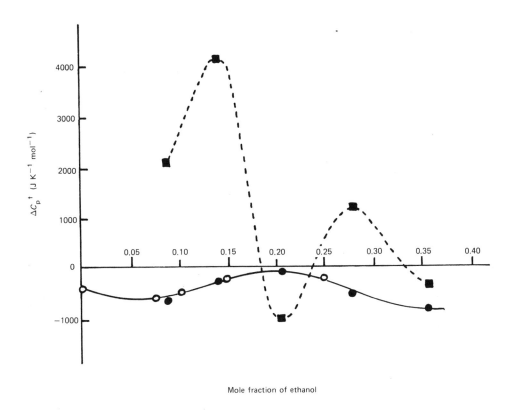

Fig. 5.30 — Relationship between ultrasonic (■) and non-ultrasonic (●) $\Delta C_p^{\ddagger}$ values ○ Lit. values.

We have also studied the effect of power on the solvolysis reaction in 30% and 50% (w/w) ethanol–water. The results (Fig. 5.31) confirm the earlier findings of Couppis and Klinzing in that there is a position of optimum power for solvolysis reactions, the position of which is related inversely to the reaction temperature.

In that the reactions discussed above, all have half lives ($t_{1/2}$) in excess of 10 seconds, their reaction rates can be determined using conventional techniques — e.g. titrimetry, conductivity, spectroscopy, etc. For faster reactions it is necessary to use the rapid mixing techniques (e.g. stopped, continuous and accelerated flow) which allow the time scale to be reduced to 10^{-3} seconds, or the pertubation methods which allow a reduction of $t_{1/2}$ to 10^{-6} seconds. There are two types of perturbation method, namely the Large Perturbation Method (e.g. flash photolysis, pulse radiolysis and shock tube) and the Small Perturbation Method (e.g. temperature, pressure and electric field jump). In general the Small Perturbation Method is also referred to

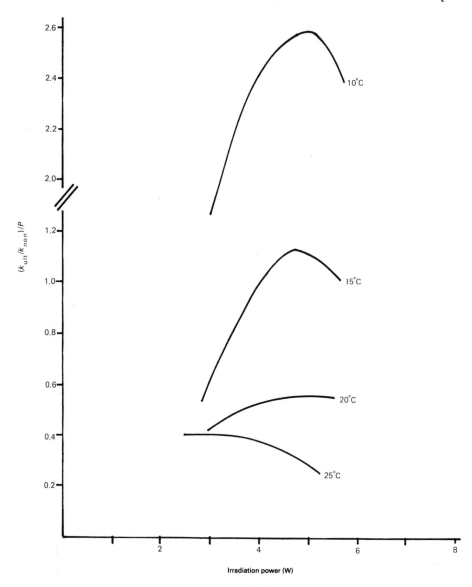

Fig. 5.31 — Relationship between race enhancement, k_{ult}/k_{non}, and irradiation power (I) for the solvolysis of 2-chloro-2-methyl propane in 50% EtOH–H_2O at various temperatures.

as the Chemical Relaxation Method since it operates by subjecting a system, already in chemical equilibrium, to either a thermal, mechanical or electrical shock, and monitoring, using a fast recording device, the rate at which the system approaches the new, or reestablishes the old equilibrium position.

We have already suggested that the passage of an ultrasonic wave through a solution produces temperature and pressure fluctuations within the liquid and as such may disturb the equilibrium position. In fact by measuring the changes in the ultrasonic absorption with frequency, the relaxation times and rate constants for fast reactions may be deduced. In that this is a relaxation process employing low intensity high-frequency ultrasound, discussion of this technique will be reserved for Chapter 6.

REFERENCES

[1] D. Kristol, H. Klotz and R. C. Parker, *Tetrahedron Lett.*, 1981, **45**, 907.

[2] E. C. Couppis and G. E. Klinzing, *AIChE. J.*, 1974, **20**, 485.

[3] J. W. Chen and W. M. Kalback, *Ind. Eng. Chem. Fundam.*, 1967, **6**, 175.

[4] S. Folger and D. Barnes, *Ind. Eng. Chem. Fundam.*, 1968, **7**, 222.

[5] N. Moriguchi, *J. Chem. Soc. Jpn.*, **54**, 949.

[6] L. Lliboutry, *J. Chem. Phys.*, **41**, 173.

[7] W. C. Schumb and W. S. Rittner, *J. Am. Chem. Soc.*, 1940, **62**, 3416.

[8] N. Miller, *Trans. Faraday Soc.*, 1950, **46**, 546.

[9] H. Fricke, E. J. Hart and H. P. Smith, *J. Chem. Phys.*, 1938, **6**, 229.

[10] J. H. Todd, *Ultrasonics*, 1970, p. 234.

[11] K. S. Suslick, P. F. Schubert and J. W. Goodale, *Ultrasonics Symposium*, 1981, 612.

[12] A. Weissler, *J. Am. Chem. Soc.*, 1959, **81**, 1077.

[13] E. J. Hart, *Radiation Res.*, 1954, **1**, 53.

[14] M. Del. Duca, E. Jeager, M. O. Davies and F. Hovarka, *J. Acoust. Soc. Am.*, 1958, **30**, 301.

[15] M. Anbar and I. Pecht, *J. Phys. Chem.*, 1964, **68**, 352; 1964, **68**, 1460.

[16] A. V. Parke and D. Taylor, *J. Chem. Soc.*, 1956, 4442.

[17] C. H. Fischer, E. J. Hart and A. Henglein, 2J. Phys. Chem., 1986, **90**, 222.

[18] E. J. Hart, C. H. Fischer and A. Henglein, *J. Phys. Chem.*, 1986, **60**, 5989.

[19] A. Henglein, *Z. Naturforsch.*, 1984, **40b**, 100.

[20] A. Henglein and C. Korman, *Int. J. Radiat. Biol.*, 1985, **48**, 251.

[21] Keisuke Makino, Magdi Mossoba and P.Riesz, *J. Phys. Chem.*, 1983, **87**, 1369.

[22] D. Rehorek and E. G. Janzen, *J. Prakt. Chem.*, Band 326, **Heft 6**, 1984, 935.

[23] M. S. Doulah, *Ind. Eng. Chem. Fundam.*, 1979, **18**, 76.

[24] V. Griffing, *J. Chem. Phys.*, 1952, **20**, 939.

[25] D. Benson, *Mechanisms of Inorganic Reactions in Solution*, p. 204, McGraw-Hill, London, 1968.

[26] E. Jozefowicz and S. Witekowa, *Acta. Chim. Acad. Sci. Hung.*, 1972, **17**, 97.

[27] J. P. Lorimer, D. McKnight, A. Turner and T. J. Mason, *Ultrasonics International 87 Conf. Proc.*, p. 762.

[28] K. S. Suslick, J. J. Gawienowski, P. F. Schubert and H. H. Wang, *Ultrasonics*, 1984, **22**, 33.

[29] K. S. Suslick, D. A. Hammerton and R. E. Cline, *J. Am. Chem. Soc.*, 1986, **108**, 5641.

[30] D. J. Donaldson, M. D. Farrington and P. Kruus, *J. Phys. Chem.*, 1979, **83**, 3131.

[31] P. Kruus and T. J. Patraboy, *J. Phys. Chem.*, 1985, **89**, 3379.

[32] R. C. Weast, Ed., *Handbook of Chemistry and Physics,* 57th edn, Chemical Rubber Co., Cleveland 1976.
[33] E. Grunwald and S. Winstein, *J. Am. Chem. Soc.,* 1948, **60**, 846.
[34] A. H. Fainberg and S. Winstein, *J. Am. Chem. Soc.,* 1956, **68**, 2760.
[35] R. E. Robertson and S. E. Sugamori, *J. Am. Chem. Soc.,* 1969, **91**, 6254.
[36] E. M. Arnett and D. R. McKelvey, *Rec. Chem. Prog.,* 1965, **27**, 185.
[37] M. J. Blandamer, J. Burgess, R. E. Robertson and J. M. W. Scott, *Chem. Rev.,* 1982, **82**, 259.
[38] J. C. Burton, *J. Acoust. Soc. Am.,* 1948, **20**, 186.
[39] H. S. Frank and D. J. G. Ives, *J. Chem. Soc. Quart. Rev.,* 1966, **20**, 1.
[40] G. Kohnstam, *Adv. Phys. Chem.,* 1966, **5**, 121.

6

The uses of high frequency ultrasound in chemistry

In the preceding chapters we have concentrated mainly on the effects of power ultrasound on a chemical system and these effects have been attributed to cavitation. In Chapter 2 however the point has been made that when the irradiation frequency is sufficiently high (over 1 MHz in normal liquid systems) there is insufficient time in the rarefaction cycle for cavitation to be produced. These high frequency, but low intensity waves are used for the purposes of diagnosis either in medicine (fetal imaging), materials science (fault detection) or chemistry (analysis). In this chapter we will concentrate on some important aspects of the use of 'diagnostic' ultrasound in chemistry.

There are two measurements of the passage of ultrasound through a liquid medium which are used analytically, namely velocity and sound attenuation.

(a) The most straightforward measurement is that of velocity. The velocity depends entirely on the characteristics of the medium and so can be related to the physico-chemical properties of that medium.
(b) The measurement of the attenuation of the sound (i.e. rate of loss of intensity per unit length) as it passes through the medium is somewhat more difficult. These losses due to attenuation are most easily understood in terms of the distance which sound is carried through different media, thus we expect the same intensity sound vibration to travel increasing distances as we move from air < water < metal. The loss in intensity is due to the efficiency of translation of sound energy from one molecule to the next.

This loss can result from either:

(i) the dissipation of energy in modes other than perfect elastic collision, i.e. energy transfer in a direction other than 180° to the source
(ii) energy consumption in the form of the transfer of vibrational or rotational energy to the individual molecules of the medium or

(iii) losses due to the sound energy being used to break intermolecular physical bonds within the medium - often referred to as viscosity losses.

When either velocity or attenuation is quantified in terms of mathematical principles the methodologies of measurement can be used to provide a whole branch of analytical chemistry complementary to conventional techniques.

6.1   ULTRASOUND IN CHEMICAL ANALYSIS

Although we are all acquainted with the concept of the use of (high frequency) ultrasound in medical scanning, the idea of using such diagnostic ultrasound for chemical analysis is by no means as familiar. And yet it should be obvious that high frequency ultrasound, if it can be used to picture the fetus in the womb or to guide delicately subcutaneous surgical needles, must be an extremely accurate tool. In a sense ultrasonic scanning is being used in this context as a small dimension 'depth gauge' with sound being reflected in varying degrees from different phase transitions in the body (Table 6.1) [1].

Table 6.1 — Percentage of ultrasound energy reflected at various interfaces

Reflecting interface	% sound reflected
Fat–muscle	1.08
Muscle–blood	0.07
Bone–fat	48.91
Soft tissue–water	0.23
Soft tissue–air	99.90

For over thirty years a crude form of medical scanning has been used by the meat industry for the estimation of animal fat, particularly in pigs. It has the great merit of being non-invasive so that measurements can be carried out on livestock and this provides a potentially ideal method of monitoring growth patterns. The absolute precision of the technique has however been called into question after thorough studies of Suffolk ram carcases [2].

Let us now turn this idea on its head and consider how diagnostic ultrasound can be used for chemical analysis. In the simplified flow system illustrated in Fig. 6.1, an emitting transducer is placed opposite a detector across a pipe (an alternative arrangement would be a single unit which emits ultrasound and also receives the reflected signal pulse from the opposite wall of the pipe). The distance travelled by the sound pulse is fixed by the location of transducer and receiver and the frequency of the ultrasound is fixed by the type of transducer used. The velocity of the sound wave is however dependent upon the medium through which it passes. This is illustrated by considering the speed of sound through different liquids: carbon tetrachloride (940 m s^{-1}), acetic acid (1170 m s^{-1}), water (1480 m s^{-1}) and glycerol (1860 m s^{-1}). Clearly therefore any changes in the character of the liquid mixture flowing through the pipe will cause a change in sound velocity and result in a different time between emission and reception of the pulse. Tiny changes in time are easily detected and measured so that, provided a calibration curve has been obtained, this instrumentation will for example register any changes in the concentration of a particular component which could well indicate the progress of a reaction.

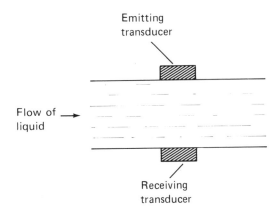

Fig. 6.1 — Non-invasive analysis of a flow system. Since the *distance* between transducers is *fixed* and *sound velocity* depends on *liquid composition* the *time* for sound to travel between transducers depends on *liquid composition*.

Even with this very simple arrangement the potential for such a method of analysis is enormous, particularly when the following points are taken into account [3].

(a) The method is totally non-invasive even to the extent that the pipework itself, being metallic, is transparent to sound.

(b) There are no moving parts involved in the measurement so that with the extremely durable modern transducers the installation should be effectively trouble free.

(c) If the calibration curve is interfaced with a computer almost instantaneous and continuous monitoring can be achieved.

(d) Unlike spectroscopic assessment any discoloration in the solution will not interfere with the ultrasonic monitoring — even opaque solutions can be handled.

This type of analysis is ideal for homogeneous mixtures and, in an ideal case, accurate measurement of the speed of sound through the medium could be used to determine chemical identity. The determination of the concentration of a dissolved species (or the solution density) is however a more common aim. Thus we can use velocity measurements for the accurate determination of the concentration of ethanol in water (Table 6.2) [4].

Table 6.2 — The determination of the velocity of ultrasound in various aqueous ethanol mixtures ($c = c_0 + At + Bt^2 + Ct^3$. c and c_0 are the sound velocities at 0°C and t°C respectively)

Mass% EtOH	c_0	A	$B \times 10^2$	$C \times 10^4$
15	1587.527	0.712	− 1.2418	− 0.58192
20	1635.701	− 0.665	− 0.38708	− 0.74471
30	1662.811	− 2.371	0.28386	0.35913
40	1610.077	− 2.356	− 0.51310	0.16872
70	1425.211	− 2.961	− 0.16765	− 0.08115
100	1230.514	− 3.515	4.18672	0.22648

The methodology is not quite so versatile when determining the concentrations of emulsions simply because not all emulsions are transparent to sound. Nevertheless for those emulsions which are transparent the relationship between concentration and velocity appears to be linear. To a lesser extent the same can be said for colloidal suspensions. One has only to turn one's attention to large-scale applications in the paint, polymer and food industries to appreciate the commercial importance of emulsions and colloidal suspensions. The determination of the stability of these mixtures is of considerable importance and the AFRC Institute of Food Research, Norwich, UK has developed a diagnostic technique for the rapid assesment of sedimentation and creaming rates for suspension and emulsion systems [5]. The methodology is very similar to that described above in that an ultrasonic velocity determination is made horizontally across a column of emulsion. The difference is that the column of material is static and it is the transducers which traverse the system in a vertical direction. Computer interfacing permits a very rapid indication of any differences in the 'density' of the column with height and therefore any separation of phases (instability) in the system.

An interesting application of the use of ultrasonic analysis for gases is to be found in the use of diagnostic ultrasound in gas liquid chromatography (glc) for the detection of trace eluents in carrier gas. The velocity of sound in a gas has been used for gas analysis for many years [6] and in glc it has been shown to have a wide range of sensitivity [7].

Recent studies have proved that this method of detection is comparable to most current glc detectors and is less subject to 'background noise' — a problem with many conventional detectors at high sensitivities [8].

It is interesting to note that ultrasound has also been used to accelerate the derivatisation of small quantities of material preliminary to glc analysis — although of course this is accomplished using power, rather than diagnostic, ultrasound and normally employs a cleaning bath [9,10].

6.2 RELAXATION PHENOMENA

6.2.1 Introduction

Whenever a sinusoidal sound wave propagates through a medium it induces oscillation of the volume occupied by the molecules and thereby increases, momentarily, the mean translational energy of the molecules. Although, in principle, this translational energy can be transferred *in toto*, by elastic collisions, to other molecules, and so increase their translational energy, in reality energy losses will occur due to two effects (a) viscosity (motion of one molecule relative to another in the medium) and (b) thermal (heat transfer from regions of high to low translational energy). It is expected therefore that the energy of the wave (I) will be attenuated as it passes through the medium. The extent of this attenuation has already been discussed in General Principles and is given by equations (2.14) and (2.17)

$$I = I_o \exp(-2\alpha d) \tag{2.14}$$

$$\alpha = \frac{2\pi^2 f^2}{\rho c^3} \left\{ \frac{4}{3}\eta_s + \eta_B + \frac{(\gamma - 1)K}{C_p} \right\} \tag{2.17}$$

Since for any given gas or liquid system η_s, η_b, γ, K and c_p are constant at constant temperature, the value, α/f^2 should be independent of the experimental frequency employed to determine α. Experimentally this is not the case and for many systems α/f^2 is observed to vary with the frequency (f). This is due to the fact that the total energy content of a medium is not restricted solely to translational energy, but is the sum of many components. For gases, the energy levels which are available in addition to translational include rotational, vibrational, and electronic, while for liquids we must also consider the possibility of conformational and structural energies. It is the coupling of the translational energy with these other energy forms which leads to the absorption of sound in excess of that deduced from eqn (2.17) and to the non-constancy of α/f^2 with increasing frequency. The occurrence of this excess absorption is most easily illustrated by considering the fate of a vibrationally excited molecule, produced as a result of the energy interchange between the translation and vibrational modes. Provided the vibrationally excited molecules can be deactivated (by inelastic collisions with other molecules) and returned to the ground state in a time period which is shorter than the period of sound oscillation, the energy will be returned to the system in phase with the sound wave, and no net loss will be observed in the sound energy per cycle. As the frequency of the sound wave increases (i.e. the time period decreases) the return of energy will become increasingly out of phase with the wave and will appear as an energy loss. Ultimately if the period of the wave is decreased sufficiently (i.e. very high frequency ultrasound is applied) a situation will be reached when the perturbation of translational energy occurs so fast that there is no time available for exchange with the other energy forms. Between these two extremes of high and low frequency there exists a condition when the frequency of the fluctuation induced by the sound wave is comparable with the time required for the energy exchange. The time lag between the excitation and de-excitation processes is observed as an acoustic relaxation. Since detailed discussions of the theoretical basis of acoustic relaxation have been published elsewhere [11,12] it is sufficient to report here that any relaxation is observable, as either an increase in the velocity–frequency curve, a peak in the $\alpha\lambda$-frequency curve, or a decrease [13,14] in the α/f^2 versus frequency curve (Fig. 6.2).

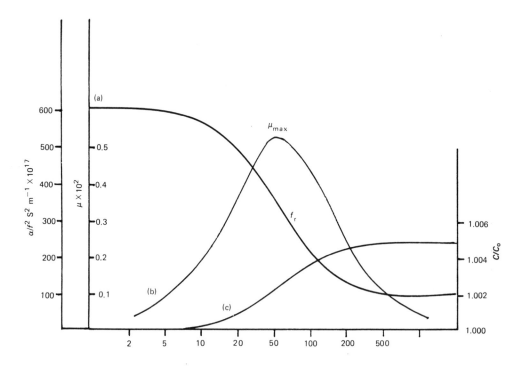

Fig. 6.2 — Variation of ultrasonic parameters with frequency for a single relaxation process
(a) α/f^2; (b) μ; (c) c.

For the latter case the experimental data may be represented by equation (6.1)

$$\frac{\alpha}{f^2} = \frac{A}{[1 + (f/f_r)^2]} + B \tag{6.1}$$

where f_r is the relaxation frequency, A is the relaxation amplitude and B is the high frequency residual absorption which is frequency independent. If more than one relaxation process (i.e. n processes) can occur, equation (6.1) is more accurately written as equation (6.2):

$$\frac{\alpha}{f^2} = \sum_{i=1}^{i=n} \frac{A_i}{[1 + (f/f_{ri})^2]} + B \tag{6.2}$$

The loss per cycle, or absorption per unit wavelength, μ, relating to the relaxation is

$$\mu = \alpha'\lambda \tag{6.3}$$

where α' is the excess absorption for the relaxation process. Thus

$$\mu = (\alpha - Bf^2)\lambda \tag{6.4}$$

or

$$\mu = \frac{Acf}{[1 + (f/f_r)^2]} \tag{6.5}$$

This curve (Fig. 6.2b) reaches a maximum value ($= \mu_m$) for μ when $f = f_r$ that is when

$$\mu_m = \tfrac{1}{2}Acf_r \tag{6.6}$$

In qualitative terms, the shape of the velocity–frequency curve may be explained in terms of increasing 'stiffness' (and hence elasticity, E) of the medium, as with increasing frequency it finds it more and more difficult to transfer the translational energy into other forms. Since velocity is related to elasticity by equation (6.7), the velocity will also increase. This change in velocity with frequency is called *velocity dispersion* and may be represented by equation (6.8), where the subscripts 0 and ∞ refer to the sound velocity, c, at low and high frequencies.

$$c = \left(\frac{E}{\rho}\right)^{\frac{1}{2}} \tag{6.7}$$

$$c^2 - c_0^2 = \frac{(2\mu_m/\pi)\, c_0 c_\infty (f/f_r)^2}{[1 + (f/f_r)^2]} \tag{6.8}$$

6.2.2 Determination of energy parameters

Figure 6.3 shows the energy level diagram for a two state equilibrium (eqn 6.9), which may be thought of as either structural, conformational or chemical, and in which state B is assumed to be the higher in energy. By determining the temperature

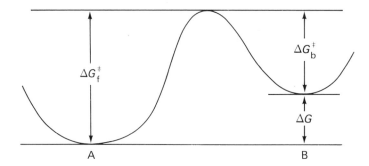

Fig. 6.3 — Energy level diagram for a two-state equilibrium.

dependences of f_r and the attenuation per wavelength (μ) it is possible to obtain information on the rate constants for the forward and back reactions, k_f and k_b, the free energy barrier, $\Delta G_b^{\ddagger}$, and the free energy difference, ΔG.

$$A \underset{k_b}{\overset{k_f}{\rightleftharpoons}} B \tag{6.9}$$

For example, if we assume for the above equilibrium that each reaction is unimolecular and takes place in an ideal solution, then the characteristic relaxation frequency, f_r, can be related to the rate constants (see below) by equation (6.10)

$$f_r = \frac{1}{2\pi\tau} = \frac{k_f + k_b}{2\pi} \tag{6.10}$$

where τ is the relaxation time of the equilibrium. Since for any reaction the rate constant, k_r, can be expressed in terms of the free energy of activation, $\Delta G^{\ddagger}$ by:

$$k_r = \frac{kT}{h} \exp\left(\frac{-\Delta G^{\ddagger}}{RT}\right) \tag{6.11}$$

then equation (6.10) may be written as

$$f_r = \frac{1}{2\pi}\frac{kT}{h}\left[\exp\left(\frac{-\Delta G_f^{\ddagger}}{RT}\right) + \exp\left(\frac{-\Delta G_b^{\ddagger}}{RT}\right)\right] \tag{6.12}$$

If we assume that the equilibrium in the above reaction lies well to the left, then we may assume that $\Delta G_f^{\ddagger} \gg \Delta G_b^{\ddagger}$ such that equation 6.12 becomes:

$$f_r = \frac{1}{2\pi} \frac{kT}{h} \exp\left(\frac{-\Delta G_b^{\ddagger}}{RT}\right)$$

(6.13)

or since $\Delta G = \Delta H - T\Delta S$,

$$f_r = \frac{1}{2\pi} \frac{kT}{h} \exp\left(\frac{\Delta S_b^{\ddagger}}{RT}\right) \exp\left(\frac{-\Delta H_b^{\ddagger}}{RT}\right)$$

(6.14)

Thus if f_r is measured over a temperature range, $\Delta H_b^{\ddagger}$ and $\Delta S_b^{\ddagger}$ may be determined from the slope and intercept of the $\ln(f_r/T)$ versus $1/T$ plot. To determine ΔH, ΔG and ΔS, use is made of equation (6.15).

$$\mu_m = \frac{\pi(\gamma - 1)R}{2C_p} \left(\frac{\Delta H}{RT}\right)^2 \exp\left(\frac{-\Delta H}{RT}\right) \exp\left(\frac{\Delta S}{R}\right)$$

(6.15)

In practice the values of C_p (molar heat capacity at constant pressure) and γ may not be known with sufficient accuracy, and it is preferable to replace them with the measured velocity of sound (eqn 6.16)

$$c^2 = \frac{(\gamma - 1)C_p}{T\theta^2}$$

(6.16)

where θ is the expansion coefficient. Substitution into equation (6.15) yields:

$$\frac{\mu_m}{c^2} \frac{\pi}{} = \frac{\pi R\theta^2}{2} \left(\frac{\Delta H}{C_p}\right)^2 \exp\left(\frac{-\Delta H}{RT}\right) \exp\left(\frac{\Delta S}{R}\right)$$

(6.17)

6.2.3 Pure liquids

Absorption studies have proved important in the investigation of the structural aspects of liquids. Basically liquids may be classified into three groups according to their sound absorption properties [13]. The first in which absorption only slightly exceeds the classical value; the second group, containing associated liquids, have α/α_{class} values of approximately 3. The excess sound absorption here is thought to be due to structural (e.g. H-bonding, see 6.2.4), rather than thermal relaxations, since the liquids have negative temperature coefficients of absorption [16,17]. The third, containing the majority of organic liquids, show large excess absorptions (α/α_{class} = 3–400) [15]. These latter large absorptions have been attributed to thermal relaxation due to the slow interchange of the internal and external energies.

As an example of the determination of the energy barriers to internal rotation in liquids consider 1,1,2-trichloroethane for which there exists a dynamic equilibrium between the three conformational forms (Fig. 6.4). Since the value of ΔH is finite, there will be a perturbation of the equilibrium on the passage of an ultrasonic wave.

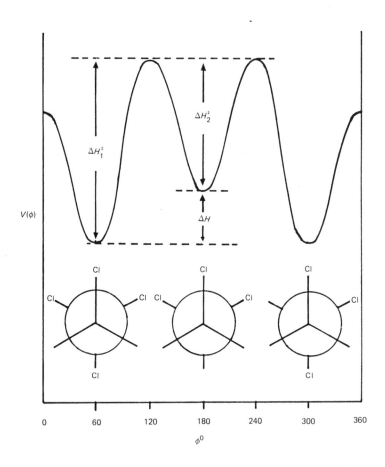

Fig. 6.4 — Potential energy $V(\phi)$ as a function of angular displacement (ϕ).

Typical graphs of the absorption per wavelength, μ, resulting from the relaxation process, as a function of frequency at various temperatures is given in Fig. 6.5. From the maxima of these curves the characteristic relaxation frequency, f_r, for each temperature may be obtained, which on applying equation (6.15) provide values of $\Delta H^{\neq}$ *and* $\Delta S^{\neq}$ of 24 kJ mol^{-1} and -11 J K^{-1} mol^{-1} respectively. By plotting $(T\mu_m/c^2)$ versus $1/T$ (eqn 6.17) a value of 9 kJ mol^{-1} for ΔH was obtained. Typical results for a selection of other substituted alkanes are given in Table 6.3, in which the values from both ultrasonic absorption and spectroscopy are compared.

For all the compounds listed above, the experimental results for the ultrasonic absorption were described by a single relaxation process (eqn 6.1) despite the fact that the compounds must have a number of different energy levels. This may imply that the observed relaxation time is an average of all the various processes.

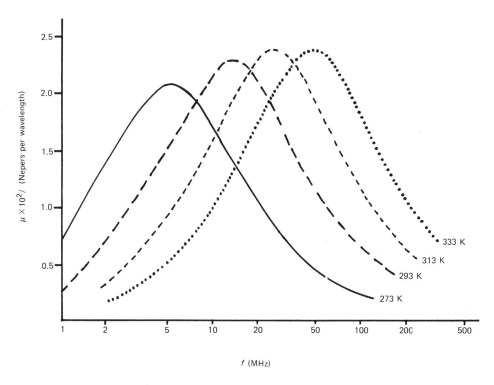

Fig. 6.5 — Ultrasonic absorption per wavelength (μ) versus frequency (f) for rotational relaxation in 1,1,2-trichloroethane

The technique of ultrasonic absorption has also been used to investigate *cis-trans* isomerisation in unsaturated aldehydes and ketones — Fig. 6.6.

It is interesting to compare the energy values in Table 6.4 for acrolein with those for crotonaldehyde and cinnamaldehyde, where in the latter substances the hydrogen atom in the X position of acrolein is replaced by a methyl and phenyl group respectively. These groups are electron donating and should strengthen the conjugation of the C_1–C_2 bond and hence increase the rotational barrier $\Delta H_b^{\neq}$ — eqn (6.14).

Table 6.4 shows that the barrier increases (21–24 kJ mol^{-1}) as the electron donating ability of the substituent increases, while the relaxation frequency decreases (176–15.7 MHz at 298 K). The increased charge separation in these molecules resulting from the inductive effect leads to stronger intramolecular forces which stabilise the higher energy *cis* confirmation.

Substitution of a similar electron donating group in the Y position should have little effect on the conjugation or the energy barrier, hence the similarity in frequencies (at a given temperature) for acrolein and methacrolein. There is, however a difference in the ΔH value. This may be due to the greater stability of the *trans*-isomer of methacrolein, when compared with acrolein, due to attraction between the carbonyl oxygen and the methyl group.

Table 6.3 — Energy parameters for rotational isomerisation of alkane derivatives

Liquid	$\Delta H_b^{\ddagger}$ (kJ mol^{-1})	ΔH (kJ mol^{-1})	
		Ultrasonic	Spectroscopic
$(CH_3)_2CHCH(CH_3)_2$	12	4	1
$CH_2BrCH_2CH_3$	15	5	2
$CH_2ClCH(CH_3)_2$	16	5	2
$CHBr_2CHBr_2$	18	4	3
$CH_2ClCCl(CH_3)_2$	19	8	0
$CH_3CHClCH_2CH_3$	19	5	3
$(CH_3)_2CHCH_2CH_3$	20	3	0
$CH_2BrCHBrCH_3$	21	4	3
$CH_2BrCBr(CH_3)_2$	23	4	3
CH_2ClCH_2Cl	24	9	4
$CBrF_2CBrF_2$	25	6	4
$CHBr_2CH_2Br$	27	7	4

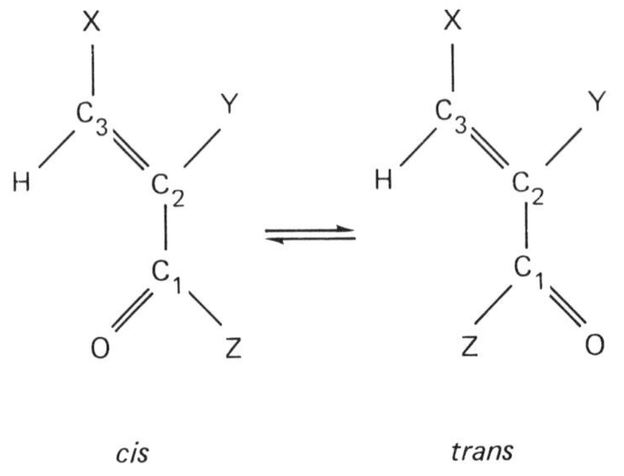

Fig. 6.6 — *Cis* and *trans* conformations of unsaturated ketones.

Although ultrasonic relaxation in liquid esters was first observed in 1936, there is a great deal of discrepancy between the results of the various workers. This is due to the fact that the characteristic relaxation frequency (f_r) tends to occur in the 100 kHz to 10 MHz range where the experimental accuracy is low. The low values are

Table 6.4 — Relaxation frequencies and energy parameters for rotational isomerisation in some unsaturated aldehydes

Liquid	T (K)	f_r (MHz)	$\Delta H_b^{\ddagger}$ (kJ mol^{-1})	$\Delta S_b^{\ddagger}$ (JK^{-1} mol^{-1})	ΔH (kJ mol^{-1})
Acrolein	248.6	29.1			
(X = Y = Z = H)	272.9	77.8	21	− 1.8	8.6
	298.4	176			
Crotonaldehyde	273.2	12.0			
(X = methyl: Y =					
Z = H)	298.2	30.3	23	− 9.0	8.1
	323.3	70.0			
Cinnamaldehyde	298.3	15.7			
(X = phenyl: Y =					
Z = H)	323.2	36.0	24	− 13	6.3
	349.0	65.2			
Methacrolein	248.4	22.2			
(Y = methyl: X =					
Z = H)	273.2	65.2	22	+ 2.0	13
	298.1	174			

explained in terms of partial double bond character in the C–O bond, due to pi
electron delocalisation, which if it occurs will lead to a substantial energy barrier for
the conversion (Fig. 6.7) and consequently low f_r values. One method of ensuring
that the relaxation frequencies are brought into the accurate experimental range
(> 5 MHz) is to work at temperatures above the normal boiling point of the liquid,
but apply a pressure to ensure the compound remains in the liquid state. This
procedure seems justified since the application of pressure does not seem to alter
significantly the relaxation frequency of ethyl ethanoate at a given temperature.-
(Table 6.5)

Typical values, obtained by the acoustic method, for various esters are given in
Table 6.6.

Conformational isomers exist for monosubstituted cyclohexanes, where the
group may be in either the equatorial or axial positions (Fig. 6.8). Since there is
greater steric interaction in the axial form, there ought to be an energy difference
between the two forms and hence an ultrasonic relaxation. Table 6.7 contains the
values of the energy parameters of three monosubstituted cyclohexanes in both the
pure and the solution state. The values of $\Delta H_b^{\ddagger}$ for chloro- and bromo-cyclohexane
agree well with the n.m.r. value of 45 kJ mol^{-1} and the infrared estimate of
63 ± 19 kJ mol^{-1}.

In deriving the relationships between ultrasonic absorption and the various
thermodynamic parameters associated with rotation, several assumptions have been
made. For example it is assumed that:

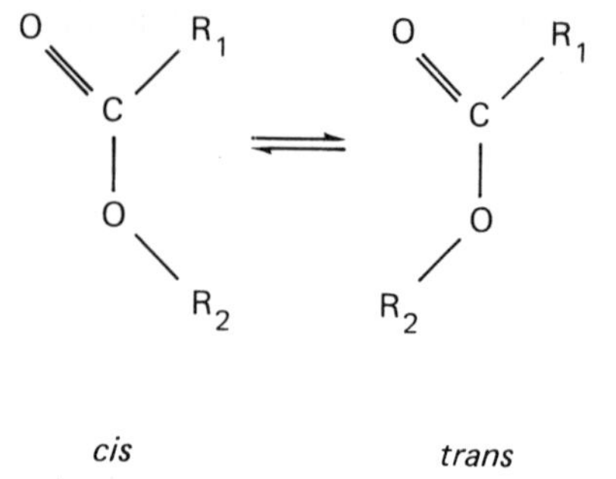

Fig. 6.7 — The two conformations of esters with the alkyl groups (R_1 and R_2) *cis* and *trans*.

Table 6.5 — Effect of pressure on the characteristic relaxation frequency of liquid ethyl ethanoate

T/K					
313	P/atm	1	381	760	1000
	f_r/MHz	22.7	23.1	22.7	23.1
353	P/atm	1	351	700	1000
	f_r/MHz	71.2	71.0	69.4	76.2

Table 6.6 — Energy parameters for rotational isomerisation of liquid esters

Ester	$\Delta H_b^{\ddagger}$ kJ mol^{-1})	$\Delta S_b^{\ddagger}$ J K^{-1} mol^{-1})	ΔH (kJ mol)	ΔS J K^{-1} mol^{-1}
Methyl methanoate	32 ± 2	-36 ± 7	$8-12$	—
Ethyl ethanoate	33 ± 2	-37 ± 5	$8-12$	—
n-Propyl methanoate	28	-28	15	7
Methyl ethanoate	25	-10	18	0
Ethyl ethanoate	18	-36	19	-4
Methyl propanoate	20	-42	21	7
Ethyl propanoate	5	-77	24	6
Ethyl butanoate	2	—	—	—

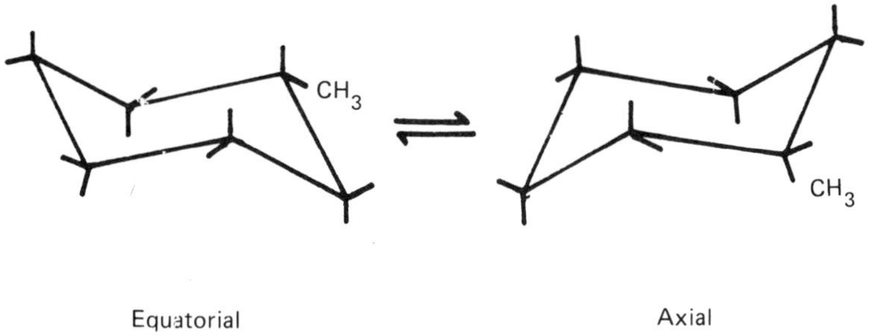

Equatorial   Axial

Fig. 6.8 — Conformations of methyl cyclohexane.

Table 6.7 — Energy parameters (in kJmol^{-1}) for inversion in substituted cyclohexanes

Liquid	Solvent	$\Delta H_b^{\neq}$	ΔH
Methylcyclohexane	Pure	43	12
	Nitrobenzene	45	15
	Xylene	46	15
	Butanol	45	—
Chlorocyclohexane	Xylene	50	—
Bromocyclohexane	Xylene	50	—

(1) The process is intramolecular and can be represented as a two-state system. Studies in different solvent systems and at various pressures (Table 6.5) show that f_r is independent of pressure and molecular environment, thus providing strong support for this assumption.

(2) The reaction is unimolecular. The fact that little dependence of f_r on concentration has been detected justifies the assumption.

(3) $\Delta G_f^{\neq} \gg \Delta G_b^{\neq}$. This assumption is justified in many cases since only one molecular relaxation has been detected by non-ultrasonic methods. Also, the plot of $\log(f_r/T)$ versus $1/T$ is linear within experimental error.

(4) $\Delta H_b^{\neq}$ and $\Delta S_b^{\neq}$ are independent of temperature. These assumptions are implicit in all methods of determing energy parameters. Since the process being studied is intramolecular, it is unlikely that this assumption is seriously in error.

(5) The dispersion of velocity in the relaxation region is negligible. This assumption is reasonable in that dispersion rarely exceeds 1%.

All the above refer to the determination of activation parameters and as such the errors and assumptions involved are not too serious. Thus the barrier heights obtained should agree with those determined in other ways. Unfortunately few data are available for comparison with the ultrasonic values, although the spectroscopic barrier heights in the substituted alkanes (Table 6.3) and cyclohexanes agree reasonably well with the acoustic methods. The ultrasonic determination of the energy differences ΔG, ΔH and ΔS yield much less satisfactory results since the assumptions are less than valid.

6.2.4 Binary liquids

Since the excess sound absorption characteristics of various binary aqueous mixtures [18–22] have attracted a great deal of attention and are well documented [23–39], we have chosen to restrict the discussion to alcohol–water mixtures which typify the behaviour of most binary systems in which solute–solvent interactions are thought to be responsible for the observed sound attenuation. The excess absorption, for a given experimental frequency varies greatly from one alcohol to another (Fig. 6.9), depends greatly on the composition of the mixture (Fig. 6.10), and decreases rapidly with rising temperature (Fig. 6.11). With a change in the experimental frequency, the magnitude of the Peak in the Sound Absorption Concentration (PSAC) decreases (Fig. 6.12).

For all binary systems the existence of a PSAC is interpreted in terms of changes in the H-bonding (i.e. solute–solvent interaction), between like and unlike molecules (AA + WW = 2AW), the transfer equilibrium being displaced by the compression of the acoustic wave. Such displacement is opposed by potential energy barriers, and the molecular translations and reorientations called upon to occur within each half-cycle introduce a relaxation which is the source of the excess absorption. As with pure liquids the combination of ultrasonic absorption and ultrasonic velocity measurements affords a means of calculating both kinetic and thermodynamic parameters for the structural change.

6.2.5 Polymers

Relaxation in polymers may be divided into processes involving segmental motion of the backbone or side groups (independent of RMM) and overall co-operative motion of the whole backbone (RMM dependent). Prior to the use of acoustic relaxation techniques the study of segmental (and side-chain) motion was restricted mainly to dielectric relaxation studies on polar molecules (dielectric studies are unable to provide information regarding motion of a non-polar molecule), with investigation of whole chain movements [40] restricted to a study of the frequency dependence of the shear viscosity of the polymer solution [41,42].

Most experimental investigations of polymers use frequencies in the range 100 kHz–500 MHz, with the attenuation determined as a function of temperature, frequency, polymer concentration and relative molar mass. Equation (6.2) is fitted, reiteratively, by computer to the experimental data [43,44] in the frequency range investigated and the best fit values of A, f_r and B are obtained by assuming that either single, double or multiple relaxation phenomena are involved.

In general the dynamic spectrum of a polymer may be divided into two parts. The first is a low frequency process, with a relaxation time ($\tau_r = 1/2\pi f_r$) which is molecular mass dependent [45–48] and an amplitude (A) which correlates with the viscosity of the solution. The second is a high frequency process which is molecular weight

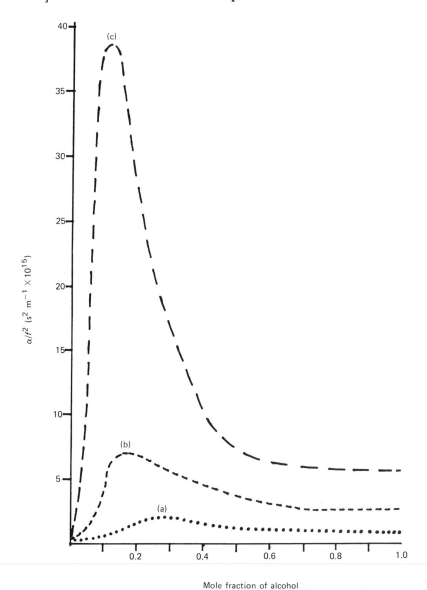

Fig. 6.9 — Sound absorption in alcohol–water mixtures. (a), ethanol; (b), isopropanol; (c), tertiary butanol.

dependent for low molecular masses and independent for higher molecular masses [49,50].

Single relaxation models [51] (i.e. one value of A and f_r) have been interpreted in terms of segmental motion of the backbone, whereas two relaxation models (A_1 and f_{r1}: A_2 and f_{r2}) have been interpreted in terms of motion of the backbone and the side

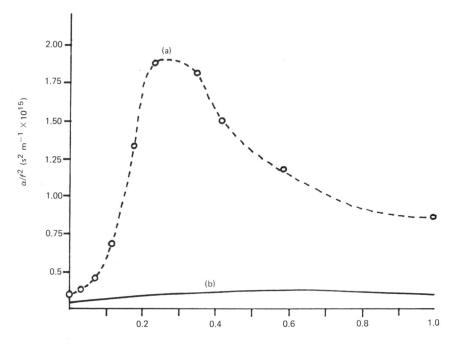

Fig. 6.10 — Sound absorption (α/f^2) as a function of ethanol composition. (a), observed absorption; (b), classical absorption.

groups. For example, the data for several polyvinyl esters [52–55] has been found to fit a double relaxation model yielding two values of A and f_r (Table 6.8).

The low frequency f_r value (3–8 MHz), being almost independent of the length of the side-chain, was interpreted as being due to motion of the backbone. The high frequency f_r value (60–150 MHz) decreased significantly with increase in the length of the side chain and was associated with reorientational motion of the side-chain.

A study of the dependence of the acoustic absorption coefficient (α) on polymer concentration (c) has in some cases yielded breaks in the α versus c curves [47,49,56,57]. These break-points have been ascribed [58] to the increased polymer–polymer interactions which occur with the onset of chain entanglement [59] in the solution.

As with pure liquids, studies of the dependence of attenuation with temperature has allowed a determination of the thermodynamic parameters (ΔG, ΔH, ΔV, ΔS) associated with the various conformational changes [60]. For example, the activation energy for polystyrene in CCl_4 [49] obtained by plotting the acoustic relaxation time against $1/T$ is in good agreement (27.3 kJ mol^{-1}) with the values obtained from dielectric studies [61] in poly(p-chlorostyrene) (21 kJ mol^{-1}) and nmr [62] measurements in the same solvent.

The energies associated with the conformational change, however, depend not only on the nature of the group attached to the backbone, but also upon the configuration (i.e. tacticity [63]) of the polymer. Certain configurations will have conformations which require lower activation energies to achieve a particular spatial

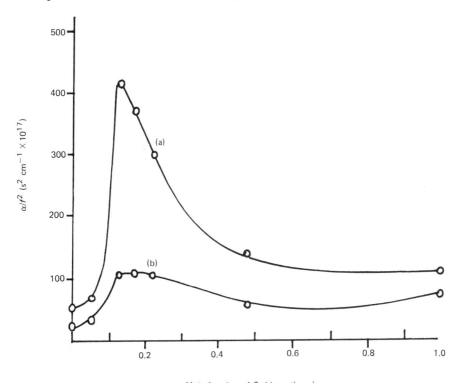

Fig. 6.11 — Sound absorption (α/f^2) of 2-chloroethanol at 0°C and 25°C ($f = 52$ MHz). (a), 0°C; (b), 25°C.

arrangement than do others. For example, the acoustic energy difference between the conformational states of poly(α-methylstyrene) (PMS), when predominantly syndiotactic, is greater than when the polymer is predominantly isotactic. This PMS value (8.3 kJ mol^{-1}) is also greater than that for the less hindered polystyrene chain (5.4 kJ mol^{-1}) chain. For poly(methyl methacrylate) [64], the energy differences for syndiotactic, atactic and isotactic are 6.3, 6.3 and 3.7 kJ mol^{-1} respectively. Changes in the tacticity of the polymer also appear to have a marked influence on both the position and amplitude of the variation of the absorption coefficient (α) with frequency.

Polyelectrolytes, combining the properties of polymers (chain flexibility) with those of electrolytes (strong electrostatic interaction) have also been investigated using ultrasonic relaxation methods [50,65]. Because of the many processes which, theoretically, could give rise to excess ultrasonic absorption, e.g. segmental motion of backbone and side groups, solvation, proton transfer and ion pair formation, caution must be exercised in assigning the relaxations to a particular process.

6.2.6 Reactions in solution
As a typical example of the use of ultrasound consider the dissociation of NH$_4$OH (eqn 6.18)

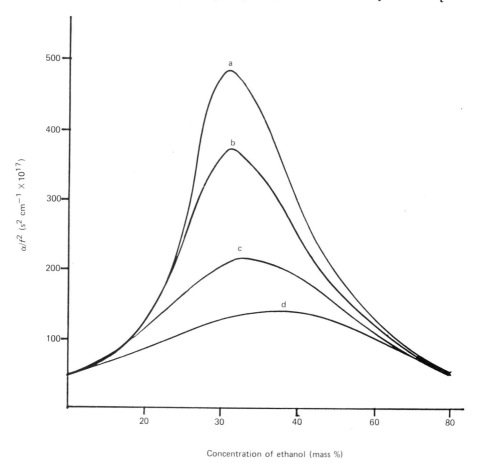

Fig. 6.12 — Variation in absorption (α/f^2) as a function of concentration for various ultrasonic frequencies ($T = 0°C$). a, 25 MHz; b, 75 MHz; c, 304 MHz; d, 1004 MHz.

$$NH_4OH \underset{k_{21}}{\overset{k_{12}}{\rightleftharpoons}} NH_4^+ + OH^- \tag{6.18}$$
$$(1-\beta)a \qquad \beta a \qquad \beta a$$

where β is the degree of dissociation and a is the initial concentration of NH_4OH.
    For such a reaction the rate equation (eqn 6.19) and equilibrium constant, K, (eqn 6.20) are respectively

$$d[NH_4^+]/dt = k_{12}[NH_4OH] - k_{21}[NH_4^+][OH^-] \tag{6.19}$$

$$K = \frac{[NH_4^+][OH^-]}{[NH_4OH]} = \frac{\beta^2 a}{(1-\beta)} = 1.5 \times 10^{-5} \text{ mol dm}^{-3} \qquad (6.20)$$

If the application of ultrasound displaces the equilibrium in such a way as to produce more NH_4^+, and if the concentration of NH_4^+ at any time t and at equilibrium are represented by x and x_e respectively, then the displacement from equilibrium at any time (Δx) may be given by

$$\Delta x = x - x_e \qquad (6.21)$$

Using this terminology equation (6.19) may be written as

$$dx/dt = k_{12}(a - x) - k_{21}x^2 \qquad (6.22)$$

In terms of the departure from equilibrium

$$d\Delta x/dt = k_{12}(a - x_e - \Delta x) - k_{21}(\Delta x + x_e)^2 \qquad (6.23)$$

At equilibrium $k_{12}(a - x_e) = k_{21}x_e^2$, such that for small departures from equilibrium $(\Delta x^2 \sim 0)$, equation (6.23) may be written as

$$d\Delta x/dt = -(k_{12} + 2k_{21}x_e)\Delta x \qquad (6.24)$$

or

$$\Delta x/\Delta x_0 = \exp\left[-(k_{12} + 2k_{21}x_e)t\right] \qquad (6.25)$$

Relaxation time, τ, is defined as the time taken for the concentration to fall to $1/e$th its initial value
i.e. $\Delta x/\Delta x_o = 1/e$. Substitution into equation (6.25) yields $1/\tau = (k_{12} + 2k_{21}x_e)$.
Since $x_e = \beta a$ then

$$1/\tau = (k_{12} + 2k_{21}\,\beta a) \qquad (6.26)$$

When $k_{21} \ll 2\beta k_{21}a$ equation (6.26) approximates to

$$1/\tau \approx 2k_{21}\,\beta a \approx 2k_{21}(Ka)^{\frac{1}{2}} \qquad (6.27)$$

Figure 6.13 shows the results for the ultrasonic absorption $(=(\Delta\alpha\lambda)/n)$ as a function of the frequency, f, for various aqueous NH_4OH solutions. ($\Delta\alpha$ is the difference in the absorption coefficient of the solution and the solvent, λ is the ultrasonic wavelength and n is the number of reactant molecules per unit volume of solvent.) For each concentration ($0.5 \, \text{mol dm}^{-3}$–$5 \, \text{mol dm}^{-3}$), the curves exhibit a maximum at a frequency, f_r, which is characteristic for the reaction (eqn 6.18)

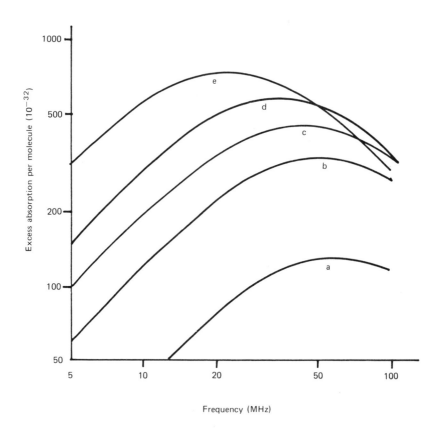

Fig. 6.13 — Frequency dependence of the excess absorption per molecule at 20°C for aqueous solutions of ammonium hydroxide. a, $5 \, \text{mol dm}^{-3}$; b, $2.5 \, \text{mol dm}^{-3}$; c, $1.5 \, \text{mol dm}^{-3}$; d, $1.0 \, \text{mol dm}^{-3}$; e, $0.5 \, \text{mol dm}^{-3}$.

Bearing in mind that τ may be represented by $1/2\pi f_r$, then from equation (6.27)

$$2\pi f_r = 2k_{21} \, (Ka)^{\frac{1}{2}} \tag{6.28}$$

The predicted linear variation of f_r with $a^{1/2}$ (eqn 6.28) can be seen in Fig. 6.14 where the slope of the line yields $k_{21} = 3 \times 10^{10} \, \text{dm}^3 \, \text{mol}^{-1} \, \text{s}^{-1}$, a value which falls within the range of rate constants expected for the diffusion controlled reaction of two ions. Other representative results are given in Table 6.8.

In some cases the reactions have been studied at different temperatures and the activation energy obtained from the appropriate Arrhenius plots. This technique has been used to determine the rate constants for such diverse reactions as the exchange

Table 6.8 — Relaxation parameters for various poly(vinyl esters) in toluene

Polymer	Temperature (K)	A_1	A_2	f_1/MHz	f_2/MHz
		$(\alpha/f^2 \times 10^{15}\ s^2\ m^{-1})$			
Poly(vinyl acetate)	273	64.5	6.2	7.9	150
	283	46.5	6.2	8.2	150
Poly(vinyl proprionate)	283	41.7	7.7	6.2	105
	293	32.6	7.7	6.5	106
Poly(vinyl butyrate)	273	30.5	15.7	3.2	57
	283	24.3	15.0	3.4	60
	293	18.2	12.0	4.5	63

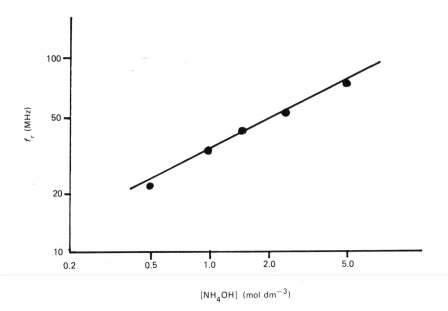

Fig. 6.14 — Double logarithmic plot of the variation of relaxation frequency (f_r) with concentration of aqueous ammonium hydroxide ($T = 20°C$).

of water molecules between the first coordination sphere of a cation and the bulk of solution and micelle formation.

Ultrasonic techniques have also been used to determine the rate of 'slow' reactions, that is of reactions where the half-life is many orders of magnitude longer than the period of the ultrasonic wave. Since the reactants and products in a reaction will have different sound velocities (eqn 6.29) and absorption coefficients (eqn 6.30), the change in these quantities with time will depend upon the rate of reaction. A typical example is the emulsion polymerisation of vinyl acetate [66], where at low conversion ($<3\%$) attenuation measurements have provided information on the growth of the particles. At higher conversions however, α is strongly influenced by the inhomogeneities in the system (e.g. the presence of foam will prevent any measurement), and ultrasonic velocity is a more sensitive tool for observing the polymerisation process.

However since neither velocity nor absorption is a simple function of the composition of the mixture (eqns 6.29 and 6.30), more conventional methods (e.g. spectroscopy) are usually preferred for following the reaction.

$$c = c_0 + \sum \frac{\Delta C_i}{\Delta a_i} a_i \tag{6.29}$$

$$\alpha = \alpha_0 + \sum \alpha_i \tag{6.30}$$

where c_o and α_o are the velocity and ultrasonic attenuation of the dispersion medium, α_i is the concentration of the ith component, $(\Delta C_i/\Delta a_i)$ is the corresponding sound-concentration coefficient and a_i are the losses due to attenuation mechanisms such as friction, thermal conduction, scattering and relaxation.

REFERENCES

[1] R. Kemberling, *Textbook of Diagnostic Ultrasonography*, Chapter 1, C. V. Mosby and Co., St. Louis, USA, 1978.
[2] K. A. Leymaster, H. J. Mersman and T. G. Jenkins, *J. Animal. Sci.*, 1985, **61**, 165.
[3] R. C. Asher, *Ultrasonics*, 1987, **25**, 17.
[4] J. Emery and S. Gasse, *Acustica*, 1979, **43**, 206.
[5] D. J. Hibberd, A. M. Howe, A. R. Mackie and M. M. Robins, *Ultrasonics International* 87, *Conference Proceedings*, 1987, Butterworths, 54.
[6] C. E. Crouthamel and H. Diehl, *Anal. Chem.*, 1948, **20**, 515.
[7] D. J. David, *Gas Chromatography Detectors*, Wiley New York, 1974, pp 144–164.
[8] K. J. Skogerboe and E. S. Yeung, *Anal. Chem.*, 1984, **56**, 2684.
[9] A. P. Thio, M. J. Kornet, H. S. I. Tan and D. H. Tompkins, *Analytical Lett*, 1979, 1009.
[10] A. R. Gholson Jr., R. H. St. Louis and H. H. Hill Jr., *J. Assoc. Off. Anal. Chem.*, 1987, **70**, 897.
[11] R. A. Pethrick, *J. Macrom. Sci. Revs, Macromol. Chem.*, 1973, **9**, 91.
[12] A. M. North and R. A. Pethrick, *International Reviews of Science, Physical*

Chemistry Series 1, ed. A. D. Buckingham and G. Allen, Butterworths, London, 1972.

[13] K. F. Herzfeld and T. A. Litontz, *Adsorption and Dispersion of Ultrasonic Waves*, Academic Press, New York, 1959.

[14] R. A. Pethrick, *Sci. Prog.*, 1970, **58**, 563.

[15] H. Nomura, S. Koda and K. Hamada, *J. Chem. Soc., Faraday Trans.* 1., 1983, **83**, 527.

[16] A. D'Aprano, I. D. Donato, G. D'Arrigo, D. Bertolini, M. Cassettari and G. Salvetti, *Mol. Phys.*, 1985, **55**, 475.

[17] J. Emery and S. Gasse, *Adv. in Molec. Relax. and Interaction Processes*, 1978, **12**, 47.

[18] J. Glinski and S. Ernst, *Pol. J. Chem.*, 1982, **56**, 339.

[19] B. Jezowska-Trzebiatowski, J. Glinski and S. Ernst, *Pol. J. Chem.*, 1984, **58**, 859.

[20] S. Nishikawa and T. Yamaguchi, *Bull. Chem. Soc. Jpn.*, 1983, **56**, 1585.

[21] S. Nishikawa and K. Kotegawa, *J. Phys. Chem.*, 1985, **89**, 2896.

[22] T. C. Bhadra and M. Basu, *Ultrasonics*, 1980, **18**, 18.

[23] N. D. T. Dale, P. A. Flavelle and P. Kruus, *Can. J. Chem.*, 1976, **54**, 355.

[24] S. Nishikawa and N. Nakao, *J. Chem. Soc., Faraday Trans. l*, 1985, **81**, 1931.

[25] C. J. Burton, *J. Acoust. Soc. Am.*, 1948, **20**, 186.

[26] G. Mikhailov and S. B. Gourevitch, *Acad. Sci., URSS*, 1956, **52**, 673.

[27] J. Thamsen, *Acustica*, 1965, **16**, 14.

[28] M. J. Blandamer, D. E. Clarke, N. J. Hidden and M. C. R. Symons, *Trans. Faraday Soc.*, 1968, **64**, 2691.

[29] M. J. Blandamer, N. J. Hidden, M. C. R. Symons and N. C. Trelvar, *Trans. Faraday Soc.*, 1968, **64**, 3242.

[30] M. J. Blandamer, *Water — A Comprehensive Treatise*, *Vol* 2, ed. F. Franks, Plenum, New York, 1973.

[31] S. Rajagopalan and S. A. Tiwari, *Acustica*, 1985, **58**, 98.

[32] Y. S. Manucharov and I. G. Mikhailov, *Sov. Phys. Acoust.*, 1977, **23**, 522.

[33] L. R. O. Storey, *J. Chem. Soc.*, 1952, 43.

[34] Y. Shindo, M. Nanbu, Y. Harada and Y. Ishida, *Acustica*, 1981, **48**, 186.

[35] V. P. Romanov and V. A. Solveyev, *Sov. Phys. Acoust.*, 1965, **11**, 68.

[36] P. Kruus, L. K. Kudrayashara, I. G. Mikhailov and V. P. Romanov, *Sov. Phys. Acoust.*, 1971, **19**, 82

[37] I. G. Mikhailov and G. B. Gurevich, *An. U.S.S.R.*, 1948, **52**, 679.

[38] M. H. Blandamer, M. J. Foster and D. Waddington, *Trans. Faraday Soc.*, 1970, **66**, 1369.

[39] J. Emery and S. Gasse, *Acustica*, 1979, **43**,206.

[40] J. D. Ferry, '*Viscoelastic Properties of Polymers*', Wiley, New York, 1971.

[41] B. H. Zimm, *J. Chem. Phys.*, 1956, **24**, 269.

[42] P. E. Rouse, *J. Chem. Phys.*, 1953, **21**, 1272.

[43] G. Schwarz, *Rev. Mod. Phys.*, 1968, **40**, 206.

[44] T. Sano and Y. Yasunga, *J. Phys. Chem.*, 1973, **77**, 2031.

[45] M. A. Cochran, A. M. North and R. A. Pethrick, *J. Chem. Soc., Faraday Trans. II*, 1974, **70**, 1274.

[46] H. Hassler and H. J. Bauer, *Kolloidn. Zh.* 1969, **230**, 194.

[47] W. Ludlow, E. Wyn-Jones and J. Rassing, *Chem. Phys. Lett.*, 1972, **13**, 477.

[48] B. Fruelich, C. Noel and L. Monneric, *Polymer*, 1979, **20**, 529.

[49] H. J. Bauer, H. Hassler and M. Immendorfer, *Faraday Discuss. Chem. Soc.*, **49**, 238.

[50] S. Kato, N. Yamauchi, H. Nomura and Y. Miyahara, *Macromolecules*, 1985, **18**, 1496.

[51] A. Juszkiewicz, A. Janowski, J. Ranachowski, S. Wartewig, P. Hauptmann and L. Alig, *Acta Polmerica*, 1985, **36**, 147.

[52] H. Nomura, S. Kato and Y. Miyahara, *J. Mat. Sci. Jpn.*, 1972, **21**, 476.

[53] H. Nomura, S. Kato and Y. Miyahara, *J. Chem. Soc. Jpn. (Chem. Ind. Chem.)*, 1972, 1241; ibid., 1973, 2398.

[54] Y. Masuda, H. Ikeda and M. Ando, *J. Mat. Sci. Jpn.*, 1971, **20**, 675.

[55] O. Funschilling, P. Lemarechal and R. Cerf, *Chem. Phys. Lett.*, 1971, **12**, 365.

[56] R. Cerf, R. Zana and S. Candau, *C. R. Seances Acad. Sci.*, 1961, **252**, 2229; ibid., 1962, **254**, 1061.

[57] P. Row-Chowdhury, *Indian J. Chem.* 1969, **7**, 692.

[58] H. Nomura, S. Kato and Y. Miyahara, *Nippon Kagaku Zasshi*, 1967, **88**, 502; ibid., 1968, **89**, 149; ibid., 1969, **90**, 250.

[59] H. R. Berger, G. Heinrich and E. Straube, *Acta Polym.*, 1986, **37**, 226.

[60] S. Nishikawa and R. Shinohara, *J. Solution Chem.*, 1986, **15**, 221.

[61] W. H. Stockmayer, H. Yu and J. E. Davis, *Polym. Preprints*, 1963, **4**, 132.

[62] D. W. McCall and F. A. Borey, *J. Polym. Sci.*, 1960, **45**, 530.

[63] J. H. Dunbar, A. M. North, R. A. Pethrick and D. B. Steinhauer, *J. Chem. Soc., Faraday Trans. II.*, 1975, **71**, 1478.

[64] C. Tondre and R. Cerf, *J. Chem. Phys.*, 1968, **65**, 1105.

[65] R. Zana, *J. Macromol. Sci. Revs., Macromol. Chem.*, 1975, **C12**, 165.

[66] P. Hauptmann, F. Dinger and R. Sauberlich, *Polymer*, 1985, **26**, 1741.

7

Ultrasonic equipment and chemical reactor design

With so many chemists turning to ultrasound as a source of energy for the acceleration or modification of chemical reactivity it becomes increasingly important that some of the electrical engineering principles which underpin the whole topic of sonochemistry are understood. The first part of this chapter is intended to provide an introduction to the principles of generation of ultrasound for the non-specialist. It is to be hoped that a chemist, armed with this information, will be in a much better position to decide on the type of equipment most appropriate for the intended laboratory application.

7.1 TYPES OF TRANSDUCER

A transducer is the name for a device capable of converting one form of energy into another, a simple example being a loudspeaker which converts electrical to sound energy. Ultrasonic transducers are designed to convert either mechanical or electrical energy into high frequency sound and there are three main types: gas driven, liquid driven and electromechanical.

7.1.1 Gas-driven transducers

These are, quite simply, whistles with high frequency output (the dog whistle is a familiar example). The history of the generation of ultrasound via whistles dates back 100 years to the work of F.Galton who was interested in establishing the threshold levels of human hearing [1]. He produced a whistle which generated sound of known frequencies and was able to determine that the normal limit of human hearing is around $16\,000$ cycles s^{-1} (16 kHz). Galton's whistle was constructed from a brass tube with an internal diameter of about two millimetres (Fig. 7.1) and operated by passing a jet of gas through an orifice into a resonating cavity. On moving the plunger the size of the cavity could be changed to alter the 'pitch' or frequency of the sound emitted.

An alternative form of gas generated ultrasound is the siren. When a solid object

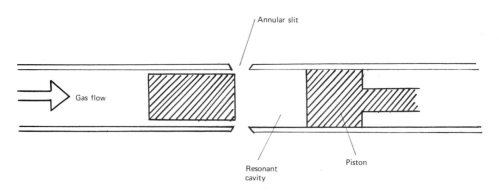

Fig. 7.1 — Galton whistle.

is passed rapidly back-and-forth across a jet of high pressure gas it interferes with the gas flow and produces sound of the same frequency at which the flow is disturbed. A siren can be designed by arranging that a gas jet impinges on the inner surface of a cylinder through which there are a series of regularly spaced perforations. When the cylinder is rotated the jet of gas emerging from a nozzle will rapidly alternate between facing a hole or the solid surface. The pitch of the sound generated by this device will depend upon the speed of rotation of the cylinder.

Neither type of transducer has any significant chemical application because it is not possible to achieve a sufficiently high intensity in airborne ultrasound by this method. Despite this, applications do exist for airborne ultrasound but the source must be very powerful — usually a sonic horn. In this way one can obtain intensities of a fraction of a watt per square centimetre in gases and at this level it is possible to use the ultrasound to break down foams, to agglomerate fine dusts and to accelerate drying processes [2].

7.1.2 Liquid-driven transducers
In essence this type of transducer can be considered as a 'liquid whistle'. If a liquid is forced rapidly from a jet across a clamped thin metal blade the blade is caused to vibrate with a frequency dependent on the flow rate (Fig. 7.2). When the vibrations are at ultrasonic frequencies the liquid suffers cavitation as it passes the blade, an analogy for which is the cavitation that is produced in water around a ship's propeller. When a mixture of immiscible liquids is forced across the blade the resulting cavitational mixing produces extremely efficient homogenisation.

7.1.3 Electromechanical transducers
The two main types of electromechanical transducers are based on either the piezoelectric or the magnetostrictive effect, the more commonly used of which are piezoelectric transducers, generally employed to power the bath and probe type sonicator systems. Although more expensive than mechanical transducers, electromechanical transducers are by far the most versatile and widely used.

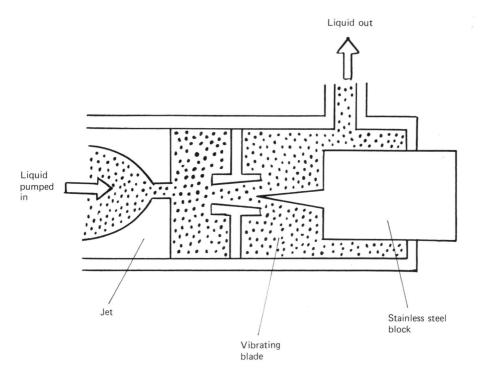

Fig. 7.2 — Whistle reactor.

Piezoelectric transducers

Currently the most common method employed for the generation and detection of ultrasound utilises the piezoelectric properties of certain crystals one of which is quartz [3]. A simplified diagram of a crystal of quartz is reproduced (Fig. 7.3) which shows three axes defined as x, y and z. If a thin section of this crystal is cut such that the large surfaces are normal to the x-axis (x-cut quartz) then the resulting section will show the following two complementary piezoelectric properties:

(a) The direct effect — when pressure is applied across the large surfaces of the section a charge is generated on each face equal in size but of opposite sign. This polarity is reversed if tension is applied across the surfaces.
(b) The inverse effect — if a charge is applied to one face of the section and an equal but opposite charge to the other face then the whole section of crystal will either expand or contract depending on the polarity of the applied charges.

Thus on applying rapidly reversing charges to a piezoelectric material, fluctuations in dimensions will be produced. This effect can be harnessed to transmit ultrasonic vibrations from the crystal section through whatever medium it might be in. However it is not possible to drive a given piece of piezoelectric crystal efficiently at every frequency. Optimum performance will only be obtained at the natural resonance frequency of the particular sample — and this depends upon its dimen-

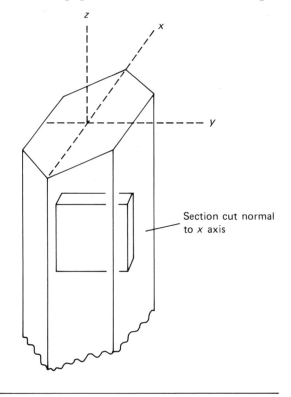

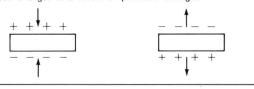

Induced charges as a result of pressure changes

Induced movement as a result of the application
of electrical potential

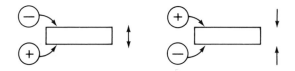

Fig. 7.3 — Simplified diagram of a quartz crystal showing X-cut.

sions. In the case of x-cut quartz a transducer of thickness 2.88 mm will have a natural frequency of 1 MHz whereas for 0.288 mm it is 10 MHz. This is why the conventional sonochemistry equipment (baths or probes) are of fixed frequency and why reports of comparative studies at different frequencies are uncommon.

There are many piezoelectric materials besides quartz. The three which are commonly used in transducers are barium titanate ($BaTiO_3$), lead metaniobate ($PbNb_2O_6$) and the mixed crystal lead zirconate titanate. These are ferroelectric compounds, i.e. they are spontaneously polarised and mechanical deformation causes a change in polarisation. These materials cannot be obtained as large single crystals and so, instead, they are ground with binders and sintered under pressure at above 1000°C to form a ceramic. The crystallites of the ceramic are then aligned by cooling from above the ferroelectric transition temperature in a magnetic field.

It is normal practice to clamp piezoelectric elements between metal blocks which serve both to protect the delicate crystalline material and to prevent it from overheating by acting as a heat sink. Usually two elements are combined so that their overall mechanical motion is additive although they are electrically opposed so that the blocks can be at earth potential. The blocks modify, by their sheer size, the nature of the ultrasonic vibrations generated. In this way a rugged reliable transducer is obtained. In Fig. 7.4 the construction of such a transducer is shown. It is generally a half-wavelength long (although multiples of this can be used) and it operates in the compressional mode. The peak-to-peak amplitudes generated by such systems are normally of the order of 10–20 micrometres [4].

The use of such different types of piezoelectric materials permits the building of ultrasonic generators of different powers and frequencies for a range of applications.

Magnetostrictive transducers

Magnetostriction refers to a change in the dimension of a suitable ferromagnetic material, e.g. nickel or iron by the application of a magnetic field and was discovered by Joule in 1847 [5]. A magnetostrictive transducer is usually in the form of a rod (or bar) acting as the magnetic core within a solenoid. Applying a varying current to the coil produces a variation in the dimensions of the bar. More recently the non-metallic, ceramic-based, ferrite materials (MFe_2O_4, M = divalent metal e.g. Ni, Zn or Pb) have become important due to their negligible eddy current losses. The advantage of a magnetostrictive transducer is that a vastly greater driving force can be applied than to a piezoelectric device. The major disadvantage is that the useful frequency range is restricted to below 100 kHz.

7.2   ULTRASONIC APPARATUS

The practising chemist has four types of laboratory ultrasonic apparatus which are commercially available. One of these, the whistle reactor, relies on mechanical generation of ultrasonic power whereas the other three — the bath, probe and cup-horn systems — are driven by electromechanical transducers. The construction of such systems is discussed below and a summary of their relative advantages (and disadvantages) in sonochemical usage are summarised in Table 7.1.

7.2.1   Whistle reactor

As noted above this type of mechanical transducer is predominantly used for homogenisation/emulsification. These devices differ markedly from the more usual bath and probe types in that they derive their power from the medium (by mechanical flow across the blade) rather than by the transfer of energy from an external source to the medium. The majority of the chemical effects observed on

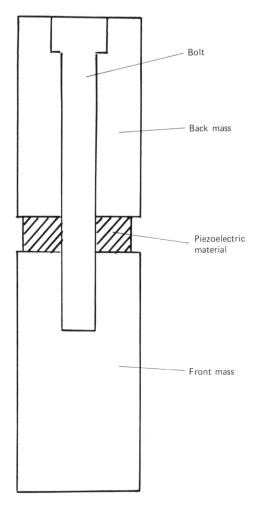

Fig. 7.4 — Construction of a modern 'sandwich' transducer.

using whistle-type transducers for the sonication of inhomogeneous reactions can be attributed mainly to the generation of very fine emulsions rather than the ultrasonic irradiation itself.

The use of the liquid whistle type of ultrasonic generators for homogenisation has increased dramatically since the Second World War. It was as far back as 1927 when Wood and Loomis reported that oil and water could be emulsified on sonication in the same beaker using quartz piezoelectric transducers [6]. However at that stage in the development of ultrasonic equipment it was quite difficult to see how transducers driven by quartz crystals could be employed on an industrial scale and so ultrasonic homogenisation remained a curiosity until 1948. In that year Janovski and Pohlmann highlighted the economic advantages to be gained from the use of a liquid whistle compared with magnetostrictive transducers of the type available at that time [7]. In

Table 7.1 — A comparison of the types of electrochemical ultrasonic apparatus

The Cleaning Bath

Advantages

   (i) The most widely available laboratory source of ultrasonic irradiation.
  (ii) Fairly even distribution of energy through reaction vessel walls.
 (iii) No special adaptation of reaction vessels required.

Disadvantages

   (i) Reduced power compared with probe system.
  (ii) Fixed frequency (and different frequencies depending on type).
 (iii) Poor temperature control.
 (iv) Position of reaction vessel in bath affects intensity of sonication.

The Probe

Advantages

   (i) High power available (no losses due to transfer through vessel walls).
  (ii) Probes can be tuned to give optimum performance at different powers.

Disadvantages

   (i) Fixed frequency.
  (ii) Difficult temperature control.
 (iii) Radical species may be generated at the tip.
 (iv) Tip erosion may occur leading to contamination by metallic particles.

The Cup-horn

Advantages

   (i) Better temperature control than a cleaning bath.
  (ii) Power control like a sonic horn but less intense.
 (iii) Little chance of radical formation in the reaction vessel.
 (iv) No fragmentation of metal into reaction from the horn tip.

Disadvantages

   (i) Reduced power compared with horn.
  (ii) Limited volume of reaction cell.
 (iii) Fixed frequency.

1960 a series of experiments was undertaken to compare four methods then in common usage for the emulsification of mineral oil, peanut oil and safflower oil [8]. The results proved that a homogeniser, which operated via a liquid whistle, was superior to three other types of apparatus namely a colloidal mill and two types of sonicator, one of which employed a quartz crystal and the other a barium titanate transducer. Subsequently the liquid whistle was adopted as an ideal method of homogenisation and it is this same design of instrument which has been used recently by Davidson to enhance the hydrolysis of fats and waxes (Fig. 7.2, see also Chapter 3) [9].

An obvious benefit of such a system is that it can be used for flow processing and can be installed 'on-line'. In this way large volumes can be processed as is the case in the manufacture of such items as fruit juices, tomato ketchup and mayonnaise.

7.2.2 Ultrasonic cleaning bath

This is probably the most accessible and cheapest piece of ultrasonic equipment available and it is for this reason that so many sonochemists begin their studies using cleaning baths.

The construction of a bath is very simple — it generally consists of a stainless steel tank with transducers clamped to its base (Fig 7.5). One of the basic parameters in

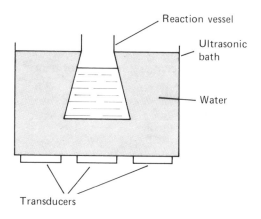

Fig. 7.5 — Schematic diagram of an ultrasonic cleaning bath.

ultrasonic engineering is power density which is defined as the electrical power into the transducer divided by the transducer radiating surface area. The low intensity (bath) system uses a power density at the transducer face of the order 1–2 W cm^{-2} for a modern piezoelectric transducer. For small baths a single transducer may be sufficient but for larger systems an array of transducers are employed to introduce high power density into the liquid contained in tanks (see for example Fig. 1.4). The frequency and power of an ultrasonic bath depends upon the type and number of transducers used in its construction. The normal method of subjecting a chemical reaction to ultrasound using a bath is simply to dip the reaction vessel in the sonicated water. The sound energy must be intense enough to penetrate the walls of the vessel and cause cavitation in the reaction. Not all laboratory cleaning baths are sufficiently powerful to achieve this and so it is important to check any bath for power before attempting to use it for sonochemistry. The easiest test to apply involves dipping a piece of ordinary household aluminium foil into the sonicated bath water (containing detergent) for about 30 seconds. Baths which are suitable for sonochemistry will perforate the foil extensively in this time.

Once the bath has been chosen the correct design for the reaction vessel must be used. For normal chemical reactions, particularly those involving heat, round-bottomed flasks are utilised but for sonochemistry in an ultrasonic bath the vessel should be one with a flat-bottom, e.g. a conical flask. The reason for this is that the energy is radiating vertically as sound waves from the base of the bath and this energy has to be transferred through the glass walls of the vessel into the reaction itself. The energy transference is much more effective when the sound impinges directly on the flat base of a conical flask rather than hitting the underside of a spherical container at an angle since when this happens more energy will be reflected away.

Another important consideration when using baths to perform sonochemical reactions is that it may be necessary to stir the mixture mechanically to achieve the maximum effect of the ultrasonic irradiation. This is particularly important when using solid–liquid mixtures where the solid is neither dispersed nor agitated throughout the reaction by sonication alone and simply sits on the base of the vessel where it is only partially available for reaction. The reason that additional stirring is so important in such cases is that it ensures the reactant powder is exposed as fully as possible to the reaction medium during sonication.

Despite the advantages gained from the use of such a simple piece of apparatus there are a number of considerations which should be borne in mind when using this method of energy input.

(a) The amount of power dissipated into the reaction from the bath is not readily quantifiable because it will depend on the size of the bath, the reaction vessel type (and thickness of its walls) and the position of the reaction vessel in the bath.
(b) Temperature control is not easy. Most cleaning baths warm up during operation especially over a prolonged period of use. This is not a problem if a heater is used to establish thermal equilibrium but can lead to inconsistent results when working around room temperature or below. However two solutions are available. Either (i) operate for very short periods during which the temperature can be assumed to remain essentially constant or (ii) circulate cooling water or add ice. However if ice is used it must be remembered that solids will alter the characteristics of sonic wave transmission. Whatever method is chosen it must be emphasised that it is the temperature inside the reaction vessel which must be monitored as this is often a few degrees above that of the bath liquid.
(c) Cleaning baths do not all operate at the same frequency and this may well affect results, particularly when attempting to reproduce those reported in the literature.

7.2.3 Direct immersion sonic horn

To overcome the disadvantages of the cleaning bath, workers in the sonochemistry field have turned to biological cell disruptors. These involve direct immersion of a metal probe into the reaction mixture. The probe — more correctly termed a sonic horn or velocity transformer — is driven by a transducer and ultrasound enters the reacting system via the probe tip. The intensity of sonication, the vibrational amplitude of the tip, can be controlled by altering the power input to the transducer and all sonicators have a power control. There is another technique for controlling power input to a system and this depends upon the type-of probe used. Most modern systems are designed to operate with a range of detachable metal probes of differing tip diameters. Probe systems are undoubtedly the most efficient method of transmitting ultrasonic energy into a reaction (Fig. 7.6).

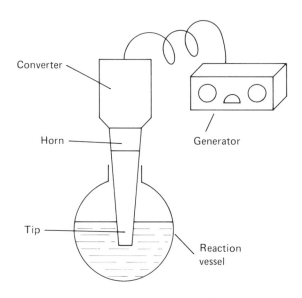

Fig. 7.6 — Schematic diagram of an ultrasonic probe system.

Horn design is a very important aspect of ultrasonic engineering [4]. The vibrational amplitude of the piezoelectric crystal itself is normally so small that the intensity of sonication attainable by direct coupling of the transducer to the chemical system is not large enough to cause cavitation. The horn acts as an amplifier for the vibration of the transducer and the precise shape of the horn will determine the gain or mechanical amplification of the vibration. It is for this reason that it is sometimes referred to as a velocity transformer. The material used for the fabrication of acoustic horns should have high dynamic fatigue strength, low acoustic loss, resistance to cavitation erosion and chemical inertness. The most suitable material by far is titanium alloy.

Horn design

Length
The wavelength of ultrasound in a material is determined by both the type of material and the frequency of the sound wave. In the case of the type of titanium alloy used for horns the wavelength for 20 kHz sound is about 26 cm and this defines absolutely the longitudinal dimension of titanium horns. Horns can be as short as half a wavelength

but should the distance between the transducer and the sample being processed need to be increased, they can be designed in multiples of half wavelengths. This can also be achieved by screwing one horn into the other thereby building up the overall length.

Shape

A selection of differently shaped horns are shown in Fig. 7.7.

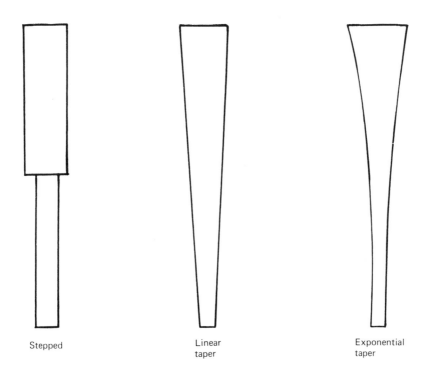

Stepped   Linear
                           taper

Exponential
taper

Fig. 7.7 — Various shapes of horns.

Stepped

If a horn is fashioned in the shape of a uniform cylinder approximately 13 cm long and subjected to ultrasonic vibrations of frequency 20 kHz at one end there will be an exactly equal vibration produced at the other end. However there will be no vibration at the midpoint of the cylinder because this is the nodal point of the wave. If the cylinder was to be reduced in diameter at its midpoint to 0.5 of the original cross-sectional area then when vibrational energy is applied to the larger end the smaller end would automatically be subject to a doubling of energy density (energy applied at larger end now emerging through half the area at the smaller end). In order to deliver this increased intensity the tip, whose vibrational frequency is fixed at 20 kHz, must vibrate at an increased amplitude — hence the horn is behaving as a 'velocity amplifier'. For a simple 'stepped' design the amplification factor will always be the ratio of the cross-sectional areas and such horns will easily accommodate gains of up to 16-fold.

The significance of the position at which the 'step' is placed should be appreciated. It is always at the nodal point of the horn because at this point there is zero vibration (i.e. no stress). If the size reduction is not precisely at this null point stress will develop at this point. Fortunately however titanium has high tensile strength and so small errors in the position of constriction can be accommodated.

Linear and exponential taper
These designs avoid the possible stress developed in stepped horns. The amplification factor for either a linear or an exponential horn is the ratio of the end diameters (not areas as with stepped).

The linear taper is the easier design to manufacture but its potential magnification is normally restricted to a factor of approximately fourfold. The exponential taper offers higher magnification factors than the linear taper. Its shape makes it more difficult to manufacture but the small diameter of the working end and its length make it particularly suited to micro applications.

For the probe system, whatever design of horn is used, a large maximum power density can be achieved at the radiating tip. This can be of the order several hundred $W \, cm^{-2}$. The working frequencies are normally of the order 20–40 kHz.

A number of probe devices are commercially available and, up to a few years ago (before the advent of sonochemistry) were referred to as cell disruptors. The majority operate at 20 kHz and utilise a wide range of different metal probes. The advantages of the probe method of energy input are threefold:

(a) Much higher ultrasonic powers can be used since energy losses during the transfer of ultrasound through the bath media and reaction vessel walls are eliminated.
(b) These devices can be tuned to give optimum performance in the reaction mixture for a range of powers.
(c) The ultrasonic intensity and size of sample to be irradiated can be matched fairly accurately for optimum effect.

However, the probe, like the bath, does suffer from the same difficulty with respect to temperature control. This problem has been alleviated to some extent in modern instruments by the incorporation of a pulse mode of operation. Quite simply this consists of a timer attached to the amplifier which switches the power to the probe on and off repeatedly. The 'off' time allows the system to cool between the pulses of sonication. The 'on' time is represented as a fraction of the total time involved in the cycle (about 1 second), i.e. 100% is continuous sonication while 50% represents 0.5 second bursts of power every 0.5 seconds.

This type of pulse operation should not be confused with 'duty cycle' as used in medical ultrasound. In medical parlance the duty cycle refers to the on/off ratio for scanning which involves the emission of pulse of extremely short duration (e.g. 10^{-5}

s) involving only a few cycles of ultrasound in the megahertz range. It is in the longer off period that the echoes are detected. By way of contrast a chemical sonicator pulse of 0.5 seconds involves 10 000 cycles at 20 kHz.

Laboratory scale reactors involving probe systems

Although it is possible to simply dip the probe into a chemical reaction contained in a standard round-bottomed flask or even a beaker, chemical reactions often require vapour tight apparatus, the slow addition of reagents or inert atmospheres. To cater for these constraints sonochemical apparatus has been developed and some of the simpler types are described below.

(a) A glass Rosett cell with flanged lid (Fig. 7.8). The design of the Rosett cell allows

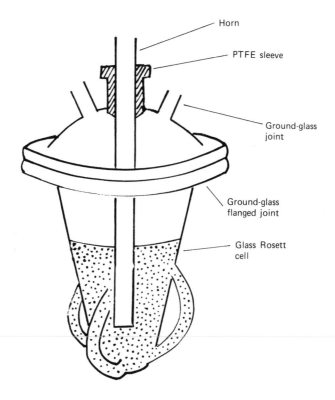

Fig. 7.8 — Rosett cell reactor.

the irradiated reaction mixture to be sonically propelled from the end of the probe around the loops of the vessel and thus provides both cooling (when the vessel is immersed in a thermostatted bath) and efficient mixing. A PTFE sleeve provides a vapour-tight fit between the probe and the glass joint [10].

As an alternative to this an ordinary reaction vessel can be adapted for sonic mixing by providing an indentation on its base to disperse the sonic waves as they are reflected from the base (Fig. 7.9).

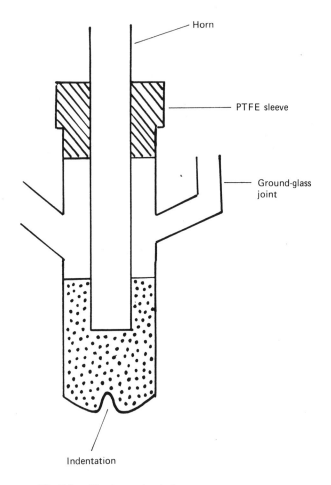

Fig. 7.9 — Simple reactor design.

(b) In situations where elevated pressures need to be used a useful reactor has been described by Suslick (Fig. 7.10) [11].

(c) A limitation of the reactors described above is that they are of the batch type. A favourite method of avoiding this is to use a flow-cell in a continuous or circulating system (Fig. 7.11).

Although the use of a probe provides several advantages over the use of baths (*vide supra*), they do suffer from a similar disadvantage - they are only capable of operating at single fixed frequency. There are however two additional problems peculiar to probe systems: (i) with direct sonication it is possible to generate radical

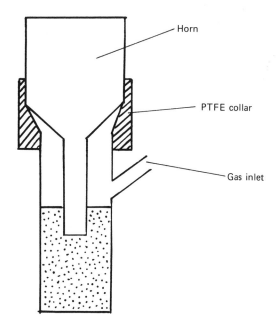

Fig. 7.10 — Pressure reactor.

species by the action of the probe tip on the solvent and (ii) with prolonged use some erosion of the tip occurs which may mean contamination of the reaction under study by small metallic particles.

The cup-horn

If the general utility of an ultrasonic bath could be combined with the controllable power of a probe a useful compromise would be available. Such a compromise system is the cup-horn device (Fig. 7.12) which can be regarded as a small bath into which can be placed the vessel containing the reaction to be sonicated. Originally designed for cell disruption, the use of such a system permits a more quantitative and reproducible study of sonochemical effects than either the cleaning bath or the probe system. The frequency is fixed but the power is tuneable and, unlike the direct immersion sonic horn, no fragments of metal are generated in the reaction itself.

The major disadvantages of the cup-horn are (i) a reduction in power compared with the direct immersion sonic horn and (ii) a restriction in size of the reaction vessel which can be placed inside the cup-horn.

7.3 LARGE-SCALE APPLICATIONS

Solutions to the problem of the scale-up of sonochemical reactions do exist but they are not as simple as the use of bigger versions of laboratory equipment. In a production situation the volumes treated will be very much larger than those considered in the laboratory and the type of process will govern the choice of reactor

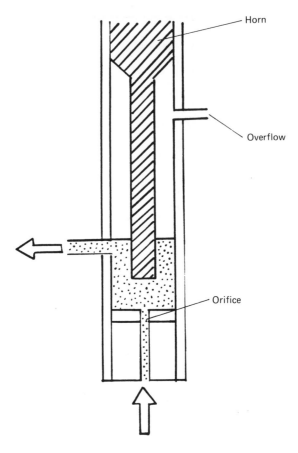

Fig. 7.11 — Flow cell.

design. It could well be that some processes would be more suited to low intensity sonication (e.g. using a bath-type reactor) whereas others may need higher intensity irradiation (via a probe-type system).

7.3.1 Whistle systems

Reference has been made already to the use of whistle reactors for homogenisation in the food industry. This system is already available for large-scale processing since it is practical, robust and durable. Although ideal for liquid processing it does seem to have inherent difficulties if used for heterogeneous solid/liquid systems. The underlying mode of action, involving as it does the flow-induced vibration of a steel blade, immediately suggests the problem of erosion by particulate matter — the sandblasting effect. Despite this apparent drawback a whistle has been used by Scott Bader Ltd, over several years without blade replacement, for the dispersion of pyrogenic silica in liquid resins at a rate of $12\,000$ dm^3 h^{-1} [12]. If indeed the blades are as durable as this would suggest it seems likely that the liquid whistle could find applications in large-scale sonochemistry involving non-corrosive materials.

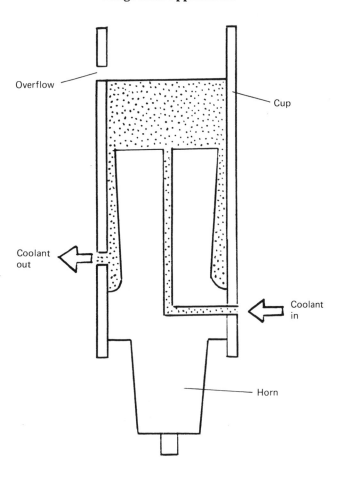

Fig. 7.12 — Cup-horn.

7.3.2 Low intensity systems

In cases where low intensity irradiation is needed, batch treatment could be as simple as using a large-scale ultrasonic cleaning bath as the reaction vessel. However the tank would need to be constructed of a material which was inert towards the chemicals involved. An appropriate grade of stainless steel might prove adequate or plastic tanks could be used. In the latter case however the transducer would need to be bonded on to a stainless or titanium plate and this assembly then bolted to the tank. A useful variant to this, and indeed one which offers greater flexibility in use is the sealed, submersible transducer assembly (Fig. 7.13). With either system some form of additional (mechanical) stirring would almost certainly be needed.

Bath-type reactors may also be used in a flow system. In this case the reacting liquids could be continuously fed into an ultrasonic tank with outflow over a weir to the next process. Such treatment could be intensified by recycling or by connecting a number of such sonicated tanks in line.

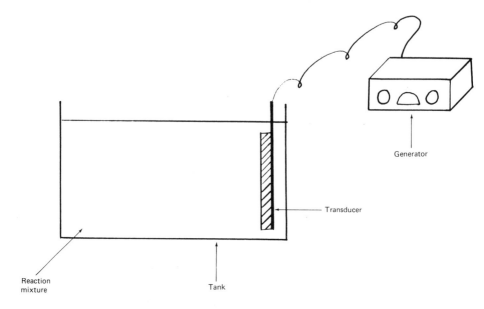

Fig. 7.13 — Submersible transducer system.

7.3.3 High intensity systems

High intensity treatment does not automatically mean the use of probes. If the application involves particle size reduction then the choice may well be the 'reverberatory ultrasonic mixing' system as developed by the Lewis Corporation in the USA (Fig. 7.14). This system can be visualised as two sonicated metal plates which enclose

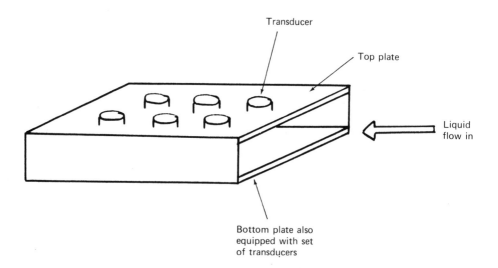

Fig. 7.14 — Reverberatory ultrasonic reactor.

a flow system. In effect the plates can be regarded as the bases of two ultrasonic baths facing toward each other and separated by only a few centimetres. Under these conditions any liquid flowing between the plates is subject to an ultrasonic intensity greater than that expected from simple doubling of a single plate intensity. The ultrasound 'reverberates' and is magnified in its effect. Already used for applications such as particle size reduction and the extraction of oil from oil shale it is clearly but a short step to using this system for chemistry. An additional benefit is that with vibrating plates the size of table tops the system can cope with very large throughputs of material.

The above is of course a flow system and we might expect that flow cells would be favourite for high intensity treatment of large volumes. The simple flow cell (Fig. 7.11) is an excellent means of processing relatively large volumes. For plant use however it is likely that a design modification would be needed which involves the coupling of a probe transducer into a flow pipe by means of a 'T' section. A number of such transducers could be employed in this manner to give extended treatment time.

Probe systems by their very nature will suffer from erosion of the horn tip and, in the long term, this could prove expensive in terms of 'down-time' for repair and replacement. A simpler solution to the problem of large-scale flow processing is the use of radial transducers surrounding the flow pipe itself (Fig. 7.15). Powerful

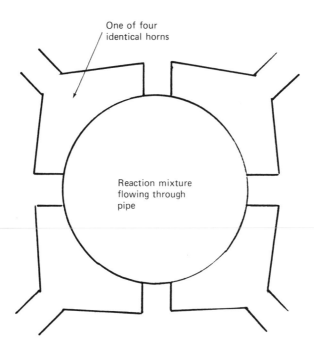

One of four
identical horns

Reaction mixture
flowing through
pipe

Fig. 7.15 — Radial transducer system.

transducers suitable for this application are currently being used to sonicate dies used for the cold drawing of metal pipes. The engineering problems associated with their construction have already been solved and so the transfer of this technology to chemical plant should be straightforward.

REFERENCES

[1] F. Galton, *Inquiries into Human Faculty and Development*, Macmillan, 1883.
[2] J. R. Frederick, *Ultrasonic Engineering*, John Wiley, 1965, Chapter 4.
[3] J. Curie and P. Curie, *Compt. Rend.*, 1880, **91**, 294.
[4] The authors are indebted to J.Perkins of Sonic Systems, Marlborough, UK for helpful discussions when writing this section.
[5] J. P. Joule, *Phil. Mag.*, 1847, **30**, 46.
[6] R. W. Wood and A. L. Loomis, *Phil. Mag.*, 1927, **4**, 417.
[7] W. Janovski and R. Pohlmann, *Z. Angew. Phys.*, 1948, **1**, 22.
[8] R. E. Singiser and H. M. Beal, *J. Am. Pharm. Assoc.* (*Scient. Ed.*), 1960, **49**, 482.
[9] R. S. Davidson, A. Safdar, J. D. Spencer and D. W. Lewis, *Ultrasonics*, 1987, **25**, 35.
[10] T. J. Mason, J. P. Lorimer, A. Moore, A. T. Turner and A. R. Harris, *Ultrasonics International* 87, *Conference Proceedings*, Butterworths, 1987, 767.
[11] K. S. Suslick and R. E. Johnson, *J. Am. Chem. Soc.*, 1984, **106**, 6856.
[12] T. J. Mason, *Laboratory Equipment Digest*, 1987, 99.

8

Miscellaneous effects of power ultrasound

In this chapter we will explore some of the effects of ultrasound which, with the exception of sonoluminescence, are not quite as much in the mainstream of sonochemical research as some of the other topics dealt with in this book. Some of these topics have a particular bearing on the development of sonochemistry with respect to processing. It is to be hoped that the topics which have been gathered together here will prove of interest and perhaps provoke research projects in new areas.

8.1 SONOLUMINESCENCE

Sonoluminescence is the name given to the weak emissions of light in an acoustically cavitating liquid. It is produced from a wide variety of liquids, both inorganic and organic, from liquid metals and even human blood plasma. The emitted light itself has been detected by a number of methods which include the naked eye [1], exposure of a photographic plate [2], the use of photomultipliers [2–5] and image intensification techniques [6, 7]. If a luminol (3-aminophthalhydrazide) solution is used (0.2 g luminol and 5 g sodium carbonate per dm^3) rather than water alone, sonoluminescence can be enhanced many times.

Despite the many original and review papers published on the topic [1–24] since it was discovered in 1933 by Marinesco and Trillat [15], the mechanism of this complex phenomenon is still not clearly understood. Of the many explanations offered for the origin of the effect, most fall into one of two broad categories, namely electrical or thermal.

8.1.1 Electrical theories

In general the basis of the electrical theories is that they propose the creation of charges during the cavitation process and that under certain conditions microdischarge occurs to produce light.

For example Frenkel [12] proposed that the creation of a cavity, initially lens-shaped, was accompanied by the production of charges on the cavity walls. By assuming that the charges were formed as a result of the non-uniform distribution of ions at the walls of bubble (i.e. opposite walls had opposite charges), he was able to determine the field strength, E, inside the bubble at the instant of formation to be:

$$ E = \left(\frac{4e}{R}\right) \sqrt{\frac{N}{d}} \qquad (8.1) $$

where e is the electronic charge, N is the number of dissociated molecules per unit volume, R is the radius of the cavity and d is the distance between the detached layers of the liquid. Subsequent analysis showed that for $d = 0.5$ nm (i.e. a value just in excess of the diameter of a water molecule, 0.2 nm) and a field strength of 60 000 V m^{-1}, electrical breakdown occurred on the creation of a cavity of radius $R = 10^{-4}$ cm. Harvey [17] for his part made the assumption that electrical breakdown, and hence microdischarge, occurred not at the instant of cavity formation, but during its collapse. Degrois and Baldo [21] in their hypothesis proposed that gas molecules, present in the liquid, were adsorbed on to the inside of the bubble surface and because of the asymmetric environment in the adsorbed configuration, each molecule was deformed giving rise to an induced dipole. It was further assumed that each adsorbed molecule was able to transfer its electronic charge in total to the bubble such that on compression the charge density increased until some critical radius was obtained. At this point electrical discharge occurred, the discharge appearing as sonoluminescence. (The dipole electrical energy, E_d, calculated for various gases is given in Table 8.1.)

Table 8.1

Gas	E_d/kJ mol^{-1}
He	2.38×10^5
Ne	22.18×10^6
Ar	5.36×10^7
Kr	41.18×10^7
Xe	11.40×10^8
H_2	1.17×10^5
N_2	22.05×10^7
O_2	34.68×10^7
Br_2	17.66×10^8
I_2	32.63×10^8

Although being able to explain the experimentally observed order of sonoluminescent intensity for the rare gases, He < Ne < Ar < Kr (i.e. the larger the E_d the greater the intensity), the E_d values are several orders of magnitude greater than would be expected even for chemisorbed species (~ 480 kJ mol^{-1}). It must therefore cast doubt on the hypothesis that sonoluminescence could occur by the mechanism suggested. Further, on the basis of the change in the magnitude of E_d, sonoluminescent intensity should vary in the order $H_2 < He < Ne < Ar < N_2 < Kr < Xe < Br_2 < I_2$. Experimentally this turns out not to be the case and the sequence observed is $H_2 < He < N_2 < O_2 < N_2 < Ar < Kr < Xe$.

It is also expected that gases with permanent dipoles will be adsorbed more readily than rare gases and should, therefore, give stronger sonoluminescence. Experiments show that sonoluminescence from solutions containing NO and NO_2 are weaker than emissions from aqueous solutions containing argon [4, 5]. This is contrary to that proposed by Degrois and Baldo's hypothesis.

Perhaps the theory which most satisfactorily accounts for the experimental observations is that by Margulis [24]. He has proposed a mechanism for the formation of electrical charges on the surface of a cavitation bubble during the detachment of small splinter bubbles (Fig. 8.1). By allowing for the flow of charge

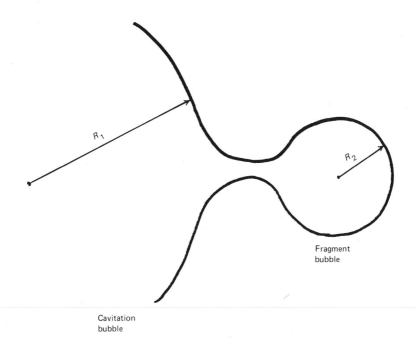

R_1

R_2

Fragment
bubble

Cavitation
bubble

Fig. 8.1 — Detachment of fragment bubble from deformed cavitation bubble.

due to the electrical conductance of the liquid, the field strength in the bubble has been calculated (1.6×10^{11} V m^{-1}) and found to be several orders of magnitude higher than the electrical breakdown strength of dry air (3×10^6 V m^{-1}) at atmospheric pressure. However, although this model more satisfactorily accounts for the experimental observations than does the previous electrical theories, the thermal theories appear to be more acceptable [19, 31].

8.1.2 Thermal theories

These include the hotspot [14, 16, 26], thermochemical [25], mechanicothermal [11, 29] and chemiluminescent [13] theories. According to the hotspot theory, sonoluminescence is merely blackbody radiation from gas within the bubble heated as the bubble undergoes a rapid adiabatic collapse. Using the ideas of Noltingk and Neppiras [14] it is a simple matter to relate the initial (T_i) and the final (T_f) bubble temperatures to the initial (R_i) and final (R_f) bubble radii by:

$$T_f = T_i (R_i/R_f)^{3(\gamma - 1)} \tag{8.2}$$

where γ is the ratio of specific heats. Srinivasan [26] concluded that for the monatomic gases, Ar and He, the equivalent blackbody temperature was equal to 11 000 K (some 5000 K higher than Gunther, Hein and Borgstedt [27]) while for the diatomic gases, O_2 and N_2, it was 8800 K. Young [28] has considered the modifying effect of the thermal diffusivity of the gas and deduced that the final collapse temperature for cavities collapsing from 10^{-4} cm to 3.33×10^{-5} cm for 20 kHz insonation ($P_A = 6$ atm) were lower, as expected, and ranged from 815 K for He to 2000 K for Xe (Fig. 8.2).

However, if Young's hypothesis is correct and can be applied generally then the predicted threshold intensity for a liquid saturated with Ar ought to be lower than one saturated with O_2 or N_2. The findings of Vaughan and Leeman [29] did not support the Young proposal and led them to consider an alternative hypothesis, the mechanicothermal theory.

The mechanicothermal theory [11] proposes that the collapse of a cavitation bubble gives rise to high temperatures, pressures and subsequent light radiation in a manner similar to that of converging shock wave. According to Bradley [30] the passage of a shock wave through a gas is 'always accompanied by a burst of visible radiation, with much of the emission spectrum attributable to continuous radiation'. Obviously shock fronts will be generated in a collapsing cavity when the converging cavity wall velocity exceeds the velocity of sound in the medium. Using the expression derived by Neppiras (eqn 8.3)

$$\dot{R}_{max} = \frac{2P_m(\gamma - 1)}{3\rho\gamma} \left[\frac{P_m(\gamma - 1)}{P\gamma} \right]^{1/(\gamma - 1)} \tag{8.3}$$

[where $\dot{R}_{max}$ is the maximum bubble wall velocity, P_m is the pressure in the liquid at the start of transient collapse and P is the pressure in the bubble at maximum size $(\sim P_v)$] and putting $\dot{R}_{max} > =$ velocity of sound, Leeman and Vaughan were able to deduce the conditions for the onset of luminescence. In general the shock wave hypothesis provided estimated values, which although slightly higher than the experimental values were in qualitative agreement (Table 8.2). The authors attributed the increase in values to their neglect of the thermal diffusivity of the gases.

Of the remaining two theories, the thermochemical theory assumes that light emission occurs as a result of the recombination of ions or radicals produced in the collapsing bubble. In the chemiluminescent theory Griffing suggested that oxidising

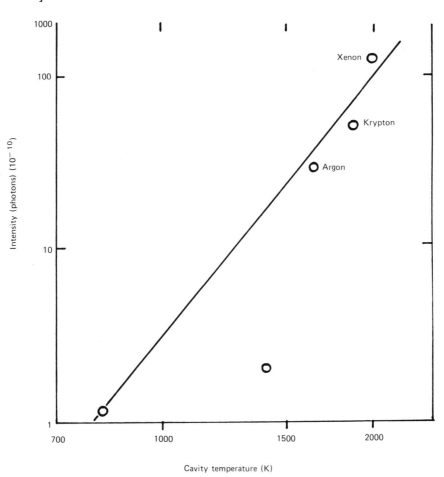

Fig. 8.2 — Sonoluminescence versus theoretical bubble temperature.

Table 8.2

Gases compared	Intensity threshold ratio	
	Experimental	Estimated
Ar/N_2	1.7	2.6
Ar/O_2	1.2	3.3

agents such as H_2O_2, formed in the bubble, would dissolve in the surrounding liquid giving rise to further reactions, some of which were of a chemiluminescent nature.

Whichever theory is accepted they all predict, in contrast to the electrical theories, that sonoluminescence emission should occur during bubble collapse. In the early 1960s a great deal of effort was devoted to investigating the relationship between sonoluminescence emission and the phase of the sound wave. The outcome was confirmation that emission always occurred near the end of a compression wave. That is not to say that stable bubbles cannot give rise to sonoluminescence. For example, in a series of experiments in which bubbles of various diameters were introduced into a viscous mixture of glycerol and water (95% volume glycerol) and then subject to ultrasound of frequency 30 kHz, Saksena and Nyborg [19] observed (Fig 8.3) that:

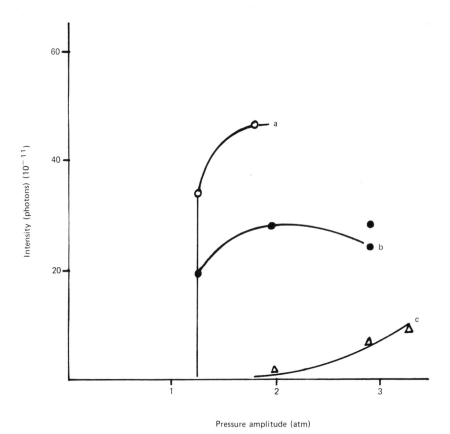

Fig. 8.3 — Variation of luminescence with pressure amplitude. a, bubble diameter of 2.5×10^{-4} m; b, bubble diameter of 5.0×10^{-4} m; c, no bubbles.

(1) There was a sharp threshold at which bubble-induced luminescence occurred.
(2) The luminescence intensity increased with increase in acoustic amplitude, but only up to a critical value.
(3) The largest luminescent intensity was observed for those bubbles whose initial diameters were closest to the resonance size (Fig. 8.3a). [Resonant size at 30 kHz was deduced from the Minneart equation (eqn 2.33).]
(4) At acoustic amplitudes greater than 2 atm generalised cavitation appeared to occur in that there was vigorous streamer activity and a high level of acoustic noise, which prior to this point had been absent. Under these conditions the sound field appeared to be of sufficient amplitude to 'create' bubbles (Fig. 8.3c), thereby giving rise to sonoluminescence — i.e. transient cavitation had commenced.
(5) The sonoluminescent flashes were seen as pulses of about 5 μsec duration and occurred with the same frequency as the sound wave.

In a further two experiments Saksena and Nyborg noted that:

(a) The addition of 1 cm^3 of allyl alcohol (a radical scavenger) to the glycerol–water mixture dramatically reduced the sonoluminescent intensity (~ 20-fold), and
(b) That the replacement of the glycerol–water mixture by a silicone fluid of similar viscosity had an almost identical (20-fold) reduction in intensity.

In that in these latter two experiments the bubble dynamics ought not to have altered significantly, Saksena and Nyborg were able to conclude that:

(1) The oscillation of stable bubbles gave rise, during the compression phase, to relatively high temperatures (~ 1800 K) within the bubble,
(2) The temperatures were inversely proportional to the vapour pressure of the medium. (In fact subsequent work showed sonoluminescence to be proportional to σ^2/P_v, where σ is the surface tension and P_v is the vapour pressure.)
(3) The emission of light was a consequence of radical recombination, the radicals produced as a result of the high bubble temperature.

There is now sufficient spectral evidence to show that sonoluminescence originates mainly from the recombination of radicals created within the high temperature and pressure environment of both transient and stable cavitation bubbles. A particular example is the emission observed by Verrall et al. [5] from water saturated with various gases. The emission (Fig. 8.4) is thought to consist of OH and H_2 band systems (from transient cavitation), overlaying a broad continuum (stable cavitation) resulting from the radiative association of the radicals (eqn 8.4)

$$H\cdot + M + \cdot OH \rightarrow H_2O + M + h\nu \qquad (8.4)$$

where M is a third body, e.g. the gas.

Verrall et al. have also used sonoluminescence to measure indirectly the inter-

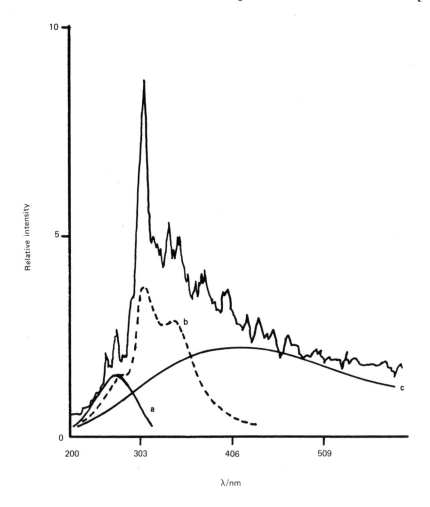

Fig. 8.4 — Sonoluminescence spectrum of argon-saturated water insonated at 333 kHz.
a, emission from H_2O; b, emission of OH; c, emission from H + OH.

cavity temperature and pressure within a collapsing bubble. This they have done by irradiating argon saturated alkali metal salt solutions at 460 kHz. Although difficult to say whether the spectra (Fig. 8.5) obtained are a result of the alkali metal ion being vaporised and then reduced, or reduced to the atomic state and subsequently vaporised, the fact that vaporisation occurs means that high temperatures exist in the cavity. This is taken as support for the hot-spot theory.

To determine the pressure (P_{max}) within the cavity, the peak half-width shift obtained acoustically was compared with shifts obtained conventionally for radiation using a Na lamp but at various experimental pressures [32, 33]. Substitution into equations (2.36) and (2.35) allowed estimates of the pressure and temperature within the cavity of 310 atmospheres and 3400 K respectively.

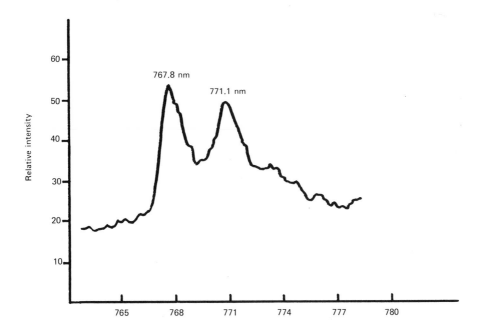

Fig. 8.5 — High resolution sonoluminescence spectrum of argon-saturated aqueous potassium
iodide solution.

Although the estimates were substantially less than those predicted theoretically (9500 K and 12 400 atm) by Noltingk and Nepparis, Verrall ascribes the differences to the former authors' neglect of the thermal conductivity of the gas. If however account was taken of the heat loss due to conduction by the gas (equation (8.5), [28]) a closer correlation between prediction (2450 K) and observation (3400 K) was obtained.

$$T_g = T_f \exp\left\{ -\frac{5\alpha}{12(2P_h/3\rho)^{1/2}R_i^{3/2}} \left[\frac{T_g^{1/2}R_f^{1/2}}{[3\ln(R_f/R_i)+1+P_A/P_h]^{1/2}} + \right.\right.$$

$$\left.\left. +\frac{T^{1/2}R_i^{1/2}}{[1+P_A/P_h]^{1/2}} \right](R_i - R_f) \right\} \tag{8.5}$$

where α = 0.92 (3R/M$^{1/2}$
      R = the gas constant, 8.314 J mol^{-1}K^{-1}
      P_h = the initial (ambient) pressure in the liquid ($\sim$ 1 atm)
      P_A = the acoustic amplitude of the sound wave

ρ = the density of the medium
T = the ambient (experimental) temperature
R_i = the initial radius of the bubble
R_f = the final radius of the bubble
T_f = the temperature in the cavity according to Nepparis
T_g = T_f corrected for heat loss due to conduction

Investigations have also been undertaken into the dependence of sonoluminescent intensity on the nature of the liquid [11, 34], the nature of the dissolved gas [4, 28, 35], the liquid temperature [4, 11, 20], the presence of impurities [19, 20, 36], ultrasonic intensity and frequency [5].

8.2 PRECIPITATION AND RELATED PROCESSES

If we define precipitation as any method of producing particulate material from liquid media then there are three such processes which have been found to be enhanced by ultrasonic irradiation — precipitation itself, crystallisation and atomisation.

Experiment has shown that power ultrasound is capable of causing both rapid and even precipitation of dissolved solids from supersaturated solutions. It has also been found possible to use ultrasound to enhance crystal growth. At first sight the processes of precipitation and crystal growth would appear to be mutually exclusive in that ultrasound is known to cause particle size reduction (*vide supra*). In several cases where particle size has been found to be increased by sonication this has been ascribed to agglomeration rather than crystal growth. Nevertheless if the right conditions are chosen it is possible to enhance either process.

8.2.1 Precipitation

The use of ultrasound has proved particularly useful in cases where extremely finely divided and uniform particles are required. Such a situation arises in the preparation of liquid dispersions of a drug for oral or subcutaneous administration where extremely small particle size gives stable suspensions of the drug plus more rapid assimilation into the body. In the 1950s a method of producing procaine penicillin was developed in which separate solutions of a procaine and a penicillin salt were mixed in an ultrasonic reactor (*ca* 100 kHz) to yield the product as a fine crystalline precipitate [34]. The particle size was smaller and much more uniform (5–15 micrometers) than that obtained using conventional mixing (10–200 micrometers).

In our own laboratories we have shown that ultrasound assists in the precipitation of magnesium carbonate in a system modelled on the commercial preparation of this material [35]. The overall reaction is shown (eqn 8.6) in which carbon dioxide is first bubbled into a 2% suspension of magnesium hydroxide to form the bicarbonate. The mixture is then filtered free of residual hydroxide before air is bubbled through to generate magnesium carbonate. The yield of carbonate by this method could be at least doubled when the second stage was subjected to sonication in an ultrasonic bath (40 kHz).

$$Mg(OH)_2 + CO_2 \rightarrow Mg(HCO_3)_2 \xrightarrow{\text{air}} Mg(CO_3)_2 \qquad (8.6)$$

On an industrial scale one important benefit of ultrasound assisted crystallisation/precipitation is that the irradiation effectively prevents encrustation of the precipitated solid on the cooling coil used for lowering the solution temperature. This makes for a more evenly distributed overall cooling rate [36].

8.2.2 Crystallisation

A major area of application of ultrasound is in metallurgical science in the casting and crystallisation of metals [37]. Sonic irradiation of a cooling melt will result in two beneficial effects — degassing (*vide supra*) and a smaller grain size. The reduction in grain size results from ultrasonic fragmentation of the orderly development of dendrites of the crystallising metal with each fission affording the opportunity for two new nucleation sites. Ultrasonic treatment of carbon steels shows a grain size reduction from 200 μm to 25–30 μm. Typical increases in ductility are 30–40% and in mechanical strength 20–30%.

The effect of sound waves on the growth of an individual crystal of zinc metal has been reported. The crystal itself was grown by the normal technique of drawing it slowly from the melt. Sonication of the growing crystal was achieved by vibrating the crystal itself as it was withdrawn. The resulting material showed large increases in material strength compared with the conventionally grown crystal as indicated by increases in critical shear stress of up to 80% (the exact figure depending on the temperature of the melt from which it was withdrawn) [38]. At 25 kHz and 50 W cm^{-2} a cylindrical zinc crystal (whose c axis is almost perpendicular to its length axis) can be changed in shape to uniform hexagonal geometry [39].

8.2.3 Atomisation

The conventional method of producing an atomised spray from a liquid is to force it at high velocity through a small aperture. (A typical domestic example being a spray-mist bottle for perfume.) Industrially such sprays are used for example in the application of various types of coating and spray drying. The disadvantage in the design of conventional equipment is that the size of the orifice restricts its usage to low viscosity liquids and such sprayers are always subject to blockage.

There are several different types of ultrasonic atomiser one of which is very similar in construction to the traditional equipment. Here the atomiser is the sonic horn itself with the liquid fed through a central tube and emerging at the vibrating tip. Although this construction is usually restricted to fairly low viscosity material the ultrasonic vibration itself does help to reduce the shear viscosity. When using sonic atomisation, it is the tip motion which causes atomisation and not the velocity of ejection through the orifice. Since the 'nozzle' is under constant agitation there is less risk of blockage with this design.

An alternative design which is much more general in that it can be applied to more viscous fluids is one where the liquid is delivered directly on to the vibrating tool. Once again there is no requirement for high velocity delivery and this design has the additional advantage that the orifice size is not important to the process and so extremely 'thick' fluids or suspensions can be handled.

In the coating industry, apart from the obvious advantages of being able to use high viscosity material with no jet blockage, sonic spraying has another great advantage over conventional methodology. Normal jet atomisation ejects particles at high velocity and some of these would collide with a surface with such high energy that 'bounce-back' might occur, particularly if there was no strong adhesion between the ejected spray and the surface. Although electrostatic methods have been employed in conventional spraying in order to reduce 'bounce-back', sonic atomisation does not produce high particle velocity and so impact adhesion is much less of a problem. The particle size of sonically atomised sprays can be accurately controlled by either the ultrasonic power or the frequency of tool vibration such that greater control of coating is possible.

Ultrasonic atomisation has also been used to produce finely particulate sprays from such materials as molten glass and metal [40]. The specific advantages to be gained from this method of producing such powders are:

(1) the particles are spherical,
(2) the particle size distribution is narrow,
(3) the particle diameter is controlled by the choice of frequency.

8.3 SEPARATION AND FILTRATION

With the need to remove suspensions of solids from liquids in both the chemical and engineering industries has come the need to improve separation technology. Conventionally membranes of various sorts have been employed ranging from the simple filter pad through semi-permeable osmotic type membranes to those which are used on a size-exclusion principle for the purification of polymeric materials. Unfortunately the conventional methodologies often lead to 'clogged' filters and, as a consequence, there will always be the need to replace filters on a regular basis. Obviously a considerable economic advantage would accrue if a non-clogging (and therefore continuously operating) filter could be developed. In this respect, the application of ultrasound provides an ideal solution to the problem. There are two specific effects of ultrasonic irradiation which can be harnessed to improve the filtration technique.

 (i) sonication will cause agglomeration of fine particles (i.e. more rapid filtration), yet
(ii) will supply sufficient vibrational energy to the system to keep the particles partly suspended and therefore leave more free 'channels' for solvent elution.

The combined influence of these effects has been successfully employed to enhance vacuum filtration of industrial mixtures such as coal slurry, which is a particularly time consuming and difficult process [41]. With the application of ultrasound to filtration, known as 'acoustic filtration', the moisture content of slurry containing 50% water can be rapidly reduced to 25% whereas conventional filtration achieves a limit of only 40%. Since coal slurry is combustible at 30% moisture content, the potential for this process is clearly enormous when applied to a continuous belt-drying process.

An improvement on the acoustic method has been developed in which an electrical potential is applied across the slurry mixture while acoustic filtration is performed. The filter itself is made the cathode while the anode, on the top of the slurry, functions as a source of attraction for the predominantly negatively charged particulate material. The additional mobility introduced by the electric charge 'electro-acoustic filtration' increases the efficiency of drying of 50% coal slurry by a further 10%.

If the filter pad could be removed altogether all problems of clogging would be completely eliminated. Just such a filtration technique is being developed to remove particulate matter from aero-engine oil by harnessing the effects of ultrasonic standing waves [42]. The device consists of two transducers set at a horizontal distance apart across a static sample of contaminated oil, or other liquid. If one transducer is activated then any particulate contaminant in the fluid is seen to collect rapidly in regions corresponding to half-wavelength distances on the axis of the ultrasonic beam. If both transducers are energised, at the same frequency, the same effect occurs. The key to this separation process is the effect on the particle 'clumps' when the transducers are operating at slightly different frequencies. Under these circumstances the standing waves (which are static when the frequencies of both transducers are identical) are seen to traverse through the medium between the transducers in a regular manner. The particles trapped at half-wavelength distances are carried with the moving 'standing wave' and thus begin to concentrate at one of the two transducers.

It is but a short step from this simple idea of particle migration to the use of the principle for particle separation. All that is now required is that the transducers are placed across a flow system and the stream beyond the ultrasonic field is cut in two (Fig. 8.6). A concentration of particulate matter will be produced in one of the two

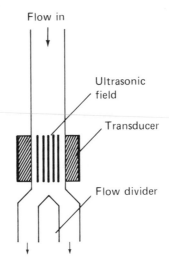

Fig. 8.6 — Particle separation in an ultrasonic field.

outflows. Recycling of these outflows may be required to achieve efficient separation.

The potential for this type of process is however far greater than simple separation/filtration. Since heavier particles are less susceptible to migration than light ones, the possibility of a sort of 'mass-spectrometric' type particle size separation exists. In essence the only modification in design that this procedure would require is that a series of cuts of the fluid (rather than two) should be made after it has passed through the ultrasonic field. Each cut would contain a slightly different spread of particle size or density.

Ultrasound mediated solute permeation through membranes has also been shown to be an important area in medicine. With the usage of power ultrasound in physiotherapy for massage and the effect that such treatment has on muscle tissue it is not too surprising to find that work was performed some years ago on ultrasonic enhancement to the absorption of medicaments through the skin. In a study of hydrocortisone uptake in swine muscle tissue via the external application of an ointment it was found that a 300% increase in efficiency could be achieved by using ultrasonic massage [43].

The potential for the medical use of ultrasound in affecting membrane permeability is not however restricted to enhanced absorption of drugs through the skin. If a medical treatment requires a constant, but low, intake of a drug it is somewhat inconvenient to ask the patient to take many regularly spaced doses in the day and night. A much easier solution exists — the drug can be encapsulated and implanted within the body so that the capsule membrane provides continuous and uniform slow release. The membrane material used for this purpose must have the specific function of slow release and this must also be at a constant rate. It would be of great benefit if, on demand and reversibly, the rate of drug release could be increased as and when the patient's condition required. Ultrasonic irradiation of the implanted capsule is easily achieved using the existing medical ultrasound equipment.

The affect of ultrasonic irradiation on interfaces has already been discussed (Chapter 3). If we turn our attention to the affect of sonication on a membrane interface it would seem reasonable to suppose that the energy of the wave may well influence permeability. This has been the interest of a few groups who have studied model *in vitro* systems, one of the simplest of which is the control of NaCl transport across a coated nylon membrane [44]. The experimental procedure is remarkably simple consisting of a small 2,12-nylon capsule containing NaCl solution immersed in a constant temperature bath filled with pure water. The progress of ionic transport across the membrane is followed by the change in conductivity of the water in the bath. The effect of insonation on the transport was assessed by monitoring conductivity when an ultrasonic probe was activated near the capsule (Fig 8.7).

In a control experiment it was determined that the rate of transport of NaCl across an untreated nylon membrane was not affected by ultrasonic irradiation. When the capsule was coated with $[CH_3(CH_2)_n]_2N^+(CH_3)_2$ Br$^-$ ($n = 11$-17) the transport at 25°C was found to be reduced by factors of between 22 and 70. Each of these coated capsules was then subjected to insonation and, in each case, the release of NaCl was accelerated. The enhancement was found to be related to the chain-length of the coating with an optimum value of 6.2-fold for $n = 15$. The rate of release

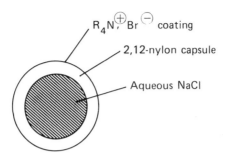

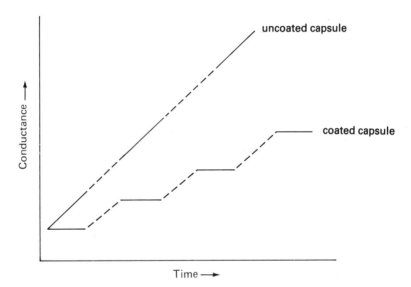

Fig. 8.7 — Ultrasonic control of membrane permeability. ——— ultrasound off; – – –
ultrasound on.

reverted to normal when the irradiation was switched off and the whole process was
found to be reproducible.

These results were related to a change in the physical state of the coating induced
by ultrasound. For the coating giving optimum enhancement the crystallisation
temperature (T_c) is 35°C and ultrasound is thought to provide sufficient energy to
cause mobilisation of the chains at 25°C and to change the state from a gel to a liquid
crystalline type. The rather smaller effect (1.7-fold) using C_{11} is attributed to its low
T_c of 8°C which would suggest that this material exists in a more liquid state at 25°C
and thus is less affected by insonation. The small effect of ultrasound (2.1-fold) on
C_{17} ($T_c = 42$°C) is attributed to the inability of ultrasound to mobilise the chains of
this material at the power used.

More recent investigations of the transport of organic compounds across simple polymer membranes with no surface coating have shown small, but significant ultrasonic enhancement [45]. The studies were carried out in aqueous solution at 25°C using 20 kHz sound and led to a 23% increase in the permeability coefficient of hydrocortisone in a cellulose membrane. For benzoic acid and poly(dimethylsiloxane) film the increase was 14%.

8.4 ELECTROCHEMISTRY

There are several aspects of ultrasound which recommend its use as an adjunct to electrochemical processes:

(a) Ultrasonic degassing will ensure that any gas evolved at an electrode surface is rapidly and efficiently removed so that gas bubble accumulation does not interfere with the passage of current.
(b) Agitation (via cavitation) at the electrode surface disturbs the diffusion layer sufficiently to stop the depletion of ions.
(c) For the same reason as above transport of ions across the electrode double layer to the electrode surface is more evenly distributed throughout the electrochemical process
(d) Cavitational collapse near an electrode surface will cause the same cleaning effect as collapse near any solid surface, i.e. the microjets produced will cause cleaning, pitting and activation of the surface — processes which are particularly beneficial in electroplating.

It is therefore somewhat surprising that there is little information on ultrasonic electrochemistry in the recent literature. Certainly in the 1950s there was considerable interest in sonically assisted electroplating [46]. Of all the plating processes studied at that time it would appear that the electrodeposition of nickel promised most in terms of ultrasonic enhancement [47]. Sonication of the plating bath was found to increase the deposition rate and to negate the fall in plating current which occurs normally due to polarisation — indeed the plating current was increased on sonication.

Similar improvements in the efficiency of chromium plating were also reported at that time [48]. Careful study revealed an improvement in the quality of the plating itself both in terms of adhesion and hardness although an increase to high current densities was accompanied by an increase in the porosity of the plate. An alternative approach is to apply ultrasound (20 kHz) directly to the cathode (low carbon steel), rather than the electrolyte solution itself [49]. The result was a better quality of plating compared with conventional methodology leading to (i) increased microhardness (10%), (ii) increased micro-crack production and (iii) better brightness.

Information on the effect of ultrasound on other electrochemical processes is not easy to find with very few papers in the chemical literature. There can be little doubt however that it will be used extensively in electrosynthesis for the selfsame reasons as given above. In Chapter 5 we discussed the effect of ultrasound on the electro-initiated cationic copolymerisation of substituted styrenes in dichloromethane [50]. The polymerisation of the individual styrenes occurred at different potentials with the composition of the copolymer dependent upon the potential used. Sonication had two effects on the reaction. First it reduced the overall potential at which polymerisation occurred, and secondly the resulting copolymer contained a larger proportion of the monomer with the higher polymerisation potential than was obtained by conventional electrolysis.

Ultrasonic irradiation has also had an influence on the properties of films produced by the electrochemical polymerisation of thiophene. In the absence of ultrasonic irradiation the films gradually became brittle as the electrolytic current density exeeded 5 mA cm^{-2}. In contrast, flexible and tough films (tensile modulus, 3.2 GPa, and strength 90 MPa) were obtained even at high current density (10 mA cm^{-2}) in the presence of ultrasound [51]. Results from cyclic voltammetry imply, not surprisingly, higher diffusion rates for the films prepared under the action of the ultrasonic field.

A preliminary study of the Kolbe electro-oxidation of cyclohexane carboxylate in methanol at a platinum electrode has revealed a distinct difference in the product distribution when the reaction was subjected to ultrasonic irradiation. Under galvanostatic conditions at 25°C two major products were formed in similar amounts: bicyclohexane produced via single electron transfer (radical combination) and methoxycyclohexane from the carbocation generated by two-electron oxidation (carbocation substitution). When the reaction was run in an ultrasonic bath but under otherwise similar conditions a striking increase in the production of cyclohexane was noted [52]. This result coupled with other changes in product ratio induced by sonication suggests that the two-electron process is more favoured under sonochemical conditions.

REFERENCES

[1] L. A. Chambers, *J. Chem. Phys.* 1937, **5**, 290.

[2] R. D. Finch, *Ultrasonics*, 1964, **1**, 87.

[3] C. Seghal, R. P. Steer, R. G. Sutherland and R. E. Verrall, *J. Chem. Phys.*, 1979, **70**, 2242.

[4] C. Seghal, R. G. Sutherland and R. E. Verrall, *J. Phys. Chem.*, 1980, **84**, 396.

[5] C. Sehgal, R. G. Sutherland and R. E. Verrall, *J. Phys. Chem.*, 1980, **84**, 388.

[6] G. T. Reynolds, A. J. Walton and S. Gruner, *Rev. Sci. Instrum.*, 1982, **53**, 1673.

[7] L. A. Crum and G. T. Reynolds, *J. Acoust. Soc. Am.*, 1985, **78**, 137.

[8] K. Negishi, *J. Phys. Soc. Jpn.*, 1961, **16**, 1450.

[9] I. E. Elpiner, *Ultrasound, Physical Chemical and Biological Effects*, Consultants Bureau, New York, 1964

[10] M. A. Margulis, *Russ. Acoustic J.*, 1969, **15**, 153.

[11] P. Jarman, *Proc. Phys. Soc. Lond.*, 1959, **73**,628; *J. Acoust. Soc. Am.*, 1960, **32**, 1459; *Sci.Progr.*, 1958, **46**, 632.

[12] Ya. I. Frenkel, *Russ. J. Phys. Chem.*, 1940, **14**, 305.

[13] V. Griffing and D. Sette, *J. Chem. Phys.*, 1952, **20**, 939; *J. Chem. Phys.*, 1955, **23**, 503; *Phys. Rev.*, 1952, **87**, 234.

[14] E. A. Neppiras, *Physics Reports*, 1980, **61**, 160; B. E. Noltingk and E. A. Neppiras, *Proc. Phys. Soc. B* (*London*), 1950, **63B**, 674; E. A. Neppiras and B. E. Noltingk, *Proc. Phys. Soc. B* (*London*), 1951, **64B**, 1032.

[15] N. Marinesco and J. J. Trillat, *C. R. Seances Acad. Sci*, 1933, **196**, 858; ibid. 1935, **200**, 548.

[16] D. Srinivasan and L. V. Holroyde, *J. Appl. Phys.*, 1961, **32**, 446; *Phys. Rev.*, 1955, **99**, 633.

[17] E. N. Harvey, *J. Am. Chem. Soc.*, 1939, **61**, 2392.

[18] K. J. Taylor and P. D. Jarman, *Aust. J. Phys.*, 1970, **23**, 319.

[19] T. K. Saksena and W. L. Nyborg, *J. Chem. Phys.*, 1970, **53**, 1722.

[20] V. P. Gunther, W. Zeil, U. Grisar and E. Heim, *Z. Electrochem.*, 1957, **61**, 188.

[21] M. Degrois and P. Baldo, *Ultrasonics*, 1974, **14**, 25.

[22] G. L. Nathanson, *Dokl. Akad. Nauk SSSR*, 1948, **59**, 83.

[23] H. G. Flynn, *Physical Acoustics*,**1B**, pp. 57–172, ed. W. P. Mason, Academic Press, New York, 1964.

[24] M. A. Margulis, *Russ. J. Phys. Chem.*, 1985, **59**, 1497.

[25] C. Sehgal, R. P. Steer, R. G. Sutherland and R. E. Verrall, *J. Phys. Chem.*, 1977, **81**, 2618.

[26] D. Srinivasan, *Sonoluminescence of Water, Diss. Abstr.*, 1955, **15**, 2249.

[27] P. Gunther, E. Hein and H. Borgstedt, *Z. Elektrochem.*, 1959, **63** 43.

[28] F. R. Young, *J. Acoust. Soc. Am.*, 1976, **60**, 100.

[29] P. W. Vaughan and S. Leeman, *Acustica*, 1959, **59**, 279.

[30] J. N. Bradley, *Shock Waves in Chemistry and Physics*, Methuen & Co, 1968, pp. 246–263.

[31] M. G. Sirotyuk, *High intensity ultrasonic fields*, ed. L. D. Rozenberg, Plenum Press, New York, 1971.

[32] Y. Chen, *Phys.Rev*, 1940, **58**, 1051; *Rev. Mod. Phys.*, 1957, **29**, 20.

[33] H. Margenau, *Phys. Rev.*, 1932, **40**, 387; ibid., 1933, **44**, 92; *Rev. Mod. Phys.*, 1959, **31**, 569.

[34] R. R. Umbdenstock, *US Patent*, **2,** 727, 892, 1955.

[35] P. R. Lutwyche, J. P. Lorimer and T. J. Mason, unpublished results, final year undergraduate chemistry project, Coventry Polytechnic, 1985.

[36] P. K. Chendke and H. S. Fogler, *Ultrasonics*, 1975, 31.

[37] O. V. Abramov, *Ultrasonics*, 1987, **25**, 73.

[38] B. Langenecker and W. H. Frandsen, *Phil. Mag.*, 1962, **7**, 2079.

[39] B. Langenecker, *Acta Cryst.*, 1964, **17**, 904.

[40] R. Pohlman, K. Heisler and M. Cichos, *Ultrasonics*, 1974, 11.

[41] N. Senapati, *Applying ultrasonics technology to enhance chemical processing*, Seminar organised by the Batelle Institute, London, 1988.

[42] J. M. Hutchinson and R. S. Sayles, *Ultrasonics International* 87 *Conference Proceedings*, Butterworths, 1987, 302.

[43] J. E. Griffin and J. C. Touchstone, *Am. J. Phys. Med.*, 1963, **42**, 77; ibid, 1972, **51**, 62

[44] Y. Okahata and H. Noguchi, *Chem. Lett.*, 1983, 1517.

[45] T. N. Julian and G. M. Zentner, *J. Pharm. Pharmacol.*, 1986, **38**, 871.

[46] B. Brown and J. E. Goodman, *High Intensity Ultrasonics*, Iliffe Books, 1965, pp. 213–220.

[47] S. R. Rich, *Proc. Am. Electroplaters' Soc.*, 1955, **42**, 137.

[48] E. Yeager, F. Hovorka and J. Dereska, *J. Acoust. Soc. Am.*, 1957, **29**, 769.

[49] E. Namgoong and J. S. Chun, *Thin Solid Films*, 1984, **120**, 153.
[50] S. Eren, L. Toppare and U. Akbulut, *Polymer Communications*, 1987, **28**, 1987.
[51] S. Osawa, M. Ito, K. Tanaka and J. Kuwano, *Synthetic Metals*, 1987, **18**, 145.
[52] D. J. Walton, A. T. Chyla and G. Smith, unpublished results from an undergraduate project, Coventry Polytechnic, 1988.

Index

Index

DATE DUE

MAY 24 '91			